AT RISK

Many disasters are a complex mix of natural hazards and human action. *At Risk* argues that the social, political and economic environment is as much a cause of disasters as the natural environment.

Published within the International Decade of Natural Disaster Reduction, this book suggests ways in which both the social and natural sciences can be analytically combined through a 'disaster pressure and release' model. Arguing that the concept of vulnerability is central to an understanding of disasters and their prevention or mitigation, the authors explore the extent and ways in which people gain access to resources.

Individual chapters apply analytical concepts to famines and drought, biological hazards, floods, coastal storms, and earthquakes, volcanoes and landslides – the 'hazards that become disasters'.

Finally, the book draws practical and policy conclusions to promote a safer environment and reduce vulnerability. It should prove of interest to students, academics and policy makers in Geography, Environmental Studies and Development Studies.

Royalties from the sales of this book are being donated to Oxfam, in support of its work in reducing people's vulnerability to hazards.

Piers Blaikie is Professor at the School of Development Studies, University of East Anglia, UK. **Terry Cannon** is Senior Lecturer in Development Studies, University of Greenwich, UK. **Ian Davis** is Managing Director of Oxford Centre for Disaster Studies, Oxford, UK. **Ben Wisner** is Professor at the School of Social Science, Hampshire College, Amherst, USA.

D0094166

AT RISK

Natural hazards, people's vulnerability,
and disasters

Piers Blaikie, Terry Cannon,
Ian Davis, and Ben Wisner

London and New York

First published 1994
by Routledge
11 New Fetter Lane, London EC4P 4EE

Simultaneously published in the USA and Canada
by Routledge
29 West 35th Street, New York, NY 10001

Reprinted 1997

Typeset in Garamond by
Florencetype Limited, Kewstoke, Avon
Printed and bound in Great Britain by
T. J. International Ltd.

British Library Cataloguing in Publication Data
A catalogue record for this book is available from the British Library.

Library of Congress Cataloguing in Publication Data
At risk: natural hazards, people's vulnerability, and disasters/
Piers Blaikie . . . [et al.].
p.   cm.
Includes bibliographical references and index.
1. Natural disasters.   I. Blaikie, Piers M.
GB5014.A82   1994
363.3′4—dc20   93–41114

ISBN 0–415–08476–8
ISBN 0–415–08477–6 (pbk)

CONTENTS

FIGURES AND TABLES

FIGURES

TABLES

PREFACE

Long before they met each other, the four authors encountered many of the hazards discussed in this book as they worked and visited in Asia, Africa, and Latin America. They shared a dissatisfaction with then prevailing views that disasters were 'natural' in a straightforward way. They shared an admiration for the ability of ordinary people to 'cope' with poverty and even calamities, and this perspective has strongly influenced the book.

The authors brought to this project complementary skills and expertise. Blaikie had written on the socio-economic background to land degradation[1] and poverty in Nepal[2] and more recently on the AIDS epidemic in Africa.[3] Cannon has incorporated teaching on hazards into his work for many years, and is active in the Famine Commission of the International Geographical Union. He has edited a collection of studies of famine,[4] and published extensively on development and environmental problems in China. Davis had spent many years studying shelter following disasters[5] and the growth of disaster vulnerability with rapid urbanization;[6] he also brought to the project years of practical work in training government officials in disaster mitigation. Wisner has been concerned with rural physical and social planning since the mid-1960s. This took the form of land-use studies,[7] research on coping with drought,[8] and work on rural energy[9] and health care delivery.[10] As the project began he was about to draw these themes together in a systematic study of 'basic-needs' approaches to development in Africa.[11]

This book has taken a long time to complete, with all the complications of multiple authorship, and the added difficulty that the four were for much of the time in three different countries. We met about six times for several days, and progressed from sketched outlines to substantial drafts at each meeting. Much paper and many electrons and floppy disks sailed back and forth between us. A great deal of ideological baggage was stripped away as a consensus view of hazards, vulnerability, and disasters emerged. We provided crash courses for each other in areas of our own expertise. The result is a fully co-authored book, although we are aware that some idiosyncrasies of style and variations in point of view may still be visible here and there.

The process had lots to recommend it, even if speed is not one of them.

Yet the book has managed to appear midway through the International Decade of Natural Disaster Reduction (IDNDR). It arrives in the context of this decade (with its overemphasis on technology and hazard management) in the hope that it will establish the vital importance of understanding vulnerability in the context of its political, social, and economic origins. The book reasserts the significance of the human factor in disasters. It tries to move beyond technocratic management to a notion of disaster mitigation that is rooted in the potential that humans have to unite, to persevere, to understand what afflicts them, and to take common action.

While the book was being written, there has been growing awareness of vulnerability to disasters, and of the range of causal factors. We welcome this groundswell of changing awareness, and appreciate the insights of many other contributors to the analysis of disasters. There are so many people to thank for their encouragement and help in the production of this book that it would be impossible to compile a list that did not offend by erroneously omitting some. So if all those people, including those affected by disaster, friends, colleagues, publishers, students, conference participants, members of NGOs, government and UN officials, will forgive us for not including their names, let us express our thanks to them in this way. In particular, we must also thank our families for their patience, for enabling our meetings and providing a great deal of help and moral support.

NOTES

1 Blaikie (1985b); Blaikie and Brookfield (1987).
2 Blaikie, Cameron, and Seddon (1977, 1980).
3 Barnett and Blaikie (1992).
4 Bohle, Cannon, Hugo, and Ibrahim (eds) (1991).
5 Davis (1978).
6 Davis (1986, 1987).
7 O'Keefe, Westgate, and Wisner (1977).
8 Wisner (1978b, 1980).
9 Wisner et al. (1987); Wisner (1987b).
10 Wisner (1976a, 1988b, 1992a); Packard, Wisner, and Bossart (eds) (1989).
11 Wisner (1988a).

Part I

FRAMEWORK AND THEORY

1

THE CHALLENGE OF DISASTERS
AND OUR APPROACH

IN AT THE DEEP END

Disasters, especially those that are connected in the minds of the public with natural hazards, are not the greatest threat to humanity. Despite the lethal reputation of earthquakes, epidemics, and famines, many more of the world's population have their lives shortened by unnoticed events, illnesses, and hunger that pass for normal existence in many parts of the world, especially (but not only) the Third World.

Occasionally earthquakes kill hundreds of thousands, and very occasionally floods kill millions at a time. But to focus on these (in the understandably humanitarian way that outsiders respond to such tragedies) is to ignore the many millions more who are not killed in such events. There is a daily and unexceptional tragedy of those whose deaths are through 'natural' causes. Under different economic and political circumstances they should have lived longer and enjoyed a better quality of life.

We feel this book is justified despite this rather artificial separation between people at risk of natural hazards and the dangers inherent in 'normal' society. Analysing disasters allows us to show why they should not be segregated from everyday living, and to show how the risks involved in disasters must be connected with the vulnerability created for many people through their normal existence. It seeks the connections between the risks people face and the reasons for their *vulnerability* to hazards. It is therefore trying to show how disasters can be perceived within the broader patterns of society, and, indeed, how analysing them in this way may provide a much more fruitful way of building policies that can help to reduce disasters and mitigate hazards.

The crucial point about understanding why disasters occur is that it is not only natural events that cause them. They are also the product of the social, political, and economic environment (as distinct from the natural environment) because of the way it structures the lives of different groups of people. There is a danger in treating disasters as something peculiar, as events which deserve their own special focus. By being separated from the social

3

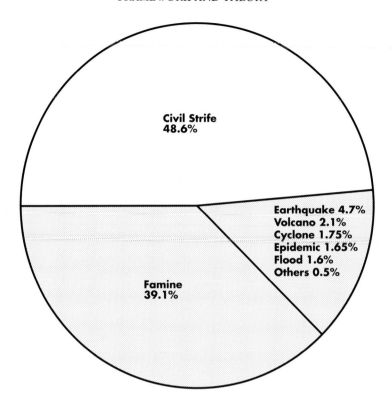

Figure 1.1 Ratio of reported deaths from disasters 1900–90

Source: Disaster History. Significant data on major disasters worldwide, Washington, DC: Office of Foreign Disaster Assistance 1990.

frameworks that influence how hazards affect people, too much emphasis in doing something about disasters is put on the natural hazards themselves, and not nearly enough on the social environment and its processes.

In this book we focus on disasters in relation to natural hazards (meteorological, geo-tectonic, biological) because that is the way most people perceive them. However, the term disaster means many different things, and some definitions include wars. Famines are often connected to wars and civil strife as well as natural events. Figure 1.1 does not distinguish between the various factors. But it is useful because it shows how few 'disaster deaths' are related to most of the natural hazards considered in this book.

Many aspects of the social environment are easily recognized: people live in adverse economic situations that lead them to inhabit parts of the world that are affected by natural hazards, be they flood plains of rivers, slopes of volcanoes, or earthquake zones. But there are many other less obvious political and economic factors that underlie the impact of hazards. These

4

involve the manner in which assets and income are distributed between different social groups, and various forms of discrimination that occur in the allocation of welfare (including relief). It is these that link our analysis of disasters that are supposedly caused mainly by natural hazards to broader patterns of society. These two aspects cannot be separated from each other: to do so risks the failure to understand the additional burden of natural hazards, and is unhelpful in both understanding disasters and doing something to prevent or mitigate them.

Many 'disasters' (as termed by outsiders or perceived by the affected people themselves) are usually a complex mix of natural hazards and human action. For example, in many regions wars are inextricably linked with famine. But the wars often either coincide with drought, or make it difficult for people to deal with drought in ways that without war they might have coped satisfactorily.

All over the world, but especially in poorer countries, vulnerable people often suffer repeated, multiple, mutually reinforcing shocks to their lives, their settlements, and their livelihoods. As often as not, the pattern of such frequent stresses, brought on by a wide variety of 'natural' trigger mechanisms, has been complicated by human action.

Continuing war in northeastern and southern Africa has made the rebuilding of lives shattered by drought virtually impossible. The deep indebtedness of many Third World countries has made the cost of reconstruction and the transition from rehabilitation to development unattainable. Rapid urbanization is putting increased numbers of people at risk. And yet the very rapid urban growth in the Third World is itself a symptom of the rural malaise of such countries.

In 'natural' disasters, a geophysical or biological event is clearly implicated in some way in causing it. Yet, even where such natural hazards appear to be directly linked to the loss of life and damage to property, the social, economic, and political origins of the disaster remain as the root causes. People's vulnerability is generated by social, economic, and political processes that influence how hazards affect people in varying ways and differing intensities.

This book is largely focused on redressing the balance in assessing the 'causes' of such disasters away from the dominant view that natural processes are most significant. The relative contributions of geophysical and biological processes, on the one hand, and social, economic, and political processes, on the other, varies from disaster to disaster. Furthermore, human activities can modify physical and biological events, sometimes at many kilometres' distance (e.g. deforestation contributing to flooding downstream) or many years later (e.g. the introduction by people of a new seed or animal, or the substitution of one form of architecture with other less safe ones). The time dimension is extremely important in another way. Social, economic, and political processes are themselves often modified by a disaster in ways that make some people more vulnerable to an extreme event in the future. The 'natural' and the 'human' are so inextricably bound together in almost all disaster situations, especially when viewed in an enlarged time and space framework, that disasters cannot be understood to be 'natural' in any straightforward way.

This is not to deny that natural events occur in which there is no social component to the disaster other than the fact that humans are in the wrong place at the wrong time. In 1986 a cloud of carbon dioxide gas bubbled up from Lake Nyos in Cameroon, killing 1,700 people in their sleep. In the complex balance of human and natural influences, this event was clearly at the 'natural' end of the spectrum of causation. The area was a long-settled, rich agricultural area. Rich and poor suffered equally. There were no differences in social or personal protection possible.

At the other extreme of the spectrum might be placed the major earthquake of 1976 in Guatemala. Like the Cameroon gas cloud, the physical shaking of the ground was a natural event. However, slum dwellers and Mayan Indians living in impoverished towns and hamlets suffered the highest mortality. The homes of the middle class were better protected and more safely sited, and recovery was easier for them. The Guatemalan poor were caught up in a vicious circle in which lack of access to means of social and personal protection made them more vulnerable to the next disaster. The social component was clear enough for a *New York Times* journalist to call the event a 'class-quake'.

Does this only mean that poor people live in flimsier houses on steeper slopes than the rich in Guatemala? Is such an obvious 'fact' as straightforward as it seems? Referring to a long history of political violence and injustice, Plant (1978) believed it was more accurate to refer to Guatemala as a 'permanent disaster'. The long history of social, economic, and political relations among the different groups in Guatemala and elsewhere has led some to argue that history 'prefigures' disaster (Hewitt 1983b). This book attempts to uncover the deeply-rooted nature of vulnerability in a way that enables us to plan for, mitigate, and perhaps prevent disasters, rather than dealing with the physical hazards as the starting point.

AUDIENCES

This book will inevitably first come to the attention of academics and students in higher education whose work interests them in disasters, development, and the Third World. We hope it will appeal to anthropologists, economists, sociologists, political scientists, and geographers, as well as others within the realm of social science. We also hope that the book will be read by natural scientists: physical geographers, geologists, oceanographers, seismologists, volcanologists, geomorphologists, hydrologists, and climatologists.

Because we see this book as being useful for action as well as study, we want to identify other groups we hope will use this book. Normally, the discussion about a book's supposed readership is found in the preface, where it seems neutral and less significant. We would rather discuss possible readers in relation to their own role in the social processes involved in making people vulnerable to hazards. By doing so we may assist in doing something to intervene in those processes to reduce that vulnerability. Such groups may include professionals involved in disaster work as an essential element in their day-to-day activity (e.g. public health workers, architects, engineers, agriculturalists, urban planners, civil servants, community activists, and politicians).

The sociologist C. Wright Mills once wrote that there are three audiences for social analysis: those with power who are aware of the consequences of their acts on others; those with power who are unaware of these consequences; and the powerless who suffer those consequences (Mills 1959). In a similar way, we identify three other broad audiences for this book: those with power who create vulnerability, sometimes without being aware of their actions; secondly, those with power who are attempting to do something about hazards, but may be unable to make it effective enough because of the failure to incorporate vulnerability analysis; and thirdly, those who are operating at the grass-roots level who suffer the consequences of disasters or are working with people to reduce their vulnerability and increase their power.

The first is the group that creates and maintains the vulnerable condition of others. Such groups include major owners of resources at international, national, and local levels (whose activities have significant effects on how and where other people live), foreign agribusiness firms, investment bankers, civil engineering contractors, and land speculators. In some cases they may be unaware of the consequences their decisions have for the vulnerability of others. This book may provoke some of them to reassess their activities, and share (in the words of the Brundtland Commission report) 'our common future' (WCED 1987b).

The second audience is extremely broad, and consists of those who attempt to address and to reduce the impact of natural hazards. It includes a variety of levels in government, and people with many interests in government activity, whose normal work is not specifically aimed at disasters as such. However, in almost every country, governments and other bodies have assumed some sort of responsibility for dealing with disasters, and this often involves measures to mitigate hazards.

At the apex of political power, leaders will take decisions on disasters possibly on advice of their senior civil servants. At this policy-formulation level, directives are developed on economic, financial, or political grounds, and will involve decisions affecting planning, agriculture, water resources, health, etc. The implementation stage will not necessarily address vulnerable conditions in relation to hazards, and indeed some policies may increase vulnerability. We hope to demonstrate that it is not enough to deal simply with the hazard threat, so that policies will be designed to reduce vulnerability and therefore disasters.

The implementation of policy extends beyond government ministries and agencies. Many voluntary agencies which have provided relief in disasters are now coming to see the need to address the pre-disaster conditions which give rise to patterns of repeated disaster and people's failure to cope. The Red Cross system is an example.[1] Following an initiative by the Swedish Red Cross (Hagman 1984), many voluntary bodies have attempted to redefine their roles in terms of 'preventing' disasters rather than just alleviating their effects. We hope our book helps to enhance their future contribution.[2]

It is also possible to find representatives of the commercial sector among those involved with vulnerability and who might be in a position to introduce mitigation measures. For instance, the logging industry can either increase risk (falling into category 'one' above) or it can work to reduce risk through measures such as selective cutting and replanting (Poore 1989). The same can be said of large-scale commercial agriculture and the mining industry, and parastatal firms such as electrical utilities (through their actions, for example, in river-basin management). A common perception that may motivate this second wide audience is that it is cheaper in the long run (in the economic, social, and political senses of the word) to prevent or mitigate disasters than to fund recovery (M. Anderson 1990).

The third group are those who are vulnerable, or who, at grass-roots level, are trying to deal with the processes that create vulnerability. We hope this book will assist organizers and activists who are part of grass-roots struggles to improve livelihoods, for instance, in the face of land deals and projects conceived by outsiders. Such locally-organized pressure groups proliferated rapidly during the 1980s and early 1990s. They are now recognized as a major force for social change in general and disaster mitigation in particular (Anderson and Woodrow 1989; Maskrey 1989; Fernando 1990).[3]

By studying disasters, much can be learned of general value about the way large-scale socio-economic systems (for instance, export trade, land tenure, urbanization) affect the life-chances of individuals and how such individuals may maintain some freedom and autonomy in the face of their vulnerability. The book may assist other lines of thinking as well. For instance, natural scientists from many disciplines are discussing the problems of uncertainty in their analysis of various natural phenomena. In situations where human actions *may* be causing catastrophic harm to natural systems at the global scale, a prudent 'precautionary science' is needed. This may apply especially in such situations where the probability of catastrophic outcome may be low but the magnitude of the catastrophe very large (Johnston and Simmonds 1991). A more conventional and optimistic view is that it is possible to 'manage the planet' if there is sufficient knowledge of all the interactions in such large-scale physical systems as the atmosphere, hydrosphere, lithosphere, asthenosphere,[4] and biosphere (W. Clark 1989). Our book will challenge these lines of thinking as well.

WHAT IS VULNERABILITY?

We have already used this word a number of times. It has a commonplace meaning: being prone to or susceptible to damage or injury. The book is an attempt to refine the definition of 'vulnerability'. To begin with we offer a simple working definition. By 'vulnerability' we mean the characteristics of a person or group in terms of their capacity to anticipate, cope with, resist, and recover from the impact of a natural hazard. It involves a combination of factors that determine the degree to which someone's life and livelihood is put at risk by a discrete and identifiable event in nature or in society.

Some groups in society are more prone than others to damage, loss, and suffering in the context of differing hazards. Key characteristics of these variations of impact include class, caste, ethnicity, gender, disability, age, or seniority. While the concept of vulnerability clearly involves varying magnitude, from high to low levels of vulnerability, for different people, we use the term to mean those who are more vulnerable. When used in this sense, the implied opposite of vulnerable is sometimes indicated by our use of the term *secure*. Other authors have contrasted vulnerability with 'capability' – the ability to protect one's community, home and family, and to re-establish one's livelihood (Anderson and Woodrow 1989).

It should also be clear that our definition of vulnerability has a time dimension built into it. Since it is damage to livelihood and not just life and property that is at issue, the more vulnerable groups are those that also find it hardest to reconstruct their livelihoods following disaster. They are therefore also more vulnerable to the effects of subsequent hazard events.

The word 'livelihood' is important in the definition. We mean by this the command an individual, family, or other social group has over an income and/or bundles of resources that can be used or exchanged to satisfy its needs. This may involve information, cultural knowledge, social networks, legal rights as well as tools, land, or other physical resources. In Chapter 3 we will develop this livelihood aspect of vulnerability in a model of *access* to opportunities which defines the ability of people to deal with the impact of the hazards to which they are exposed.[5]

Our focus on vulnerable people leads us to give secondary consideration to natural events as determinants of disasters. Normally, vulnerability is closely correlated with socio-economic position (assuming that this incorporates race, gender, age, etc.). Although we make a number of distinctions that show it to be too simplistic to explain all disasters, as a rule the poor suffer more from hazards than the rich, although poverty and vulnerability are not uniformly or invariably correlated in all cases. The key point is the significance we attach to social forms of disaster explanation.

For example, heavy rainfall may wash away the homes in wealthy hillside residential areas such as Topanga Canyon (near Los Angeles) or the Oakland–Berkeley hills (near San Francisco) as well as those of the poor in Rio de Janeiro or Belo Horizonte.[6] There are three important differences, however. Firstly, few rich people are affected if we compare the number of victims of landslides in various cities around the world. Money buys design and engineering that minimizes (but of course does not eliminate) the frequency of such events for the rich. Telecommunications and transport infrastructure facilitate warning and rescue.

Secondly, living in the hazardous canyon environment is voluntary for the rich in California, but not for the poor Brazilian or Filipino job-seeker who lives in a hillside slum. Without entering the psychological or philosophical definitions of 'voluntary' versus 'involuntary' risk-taking (Sjöberg 1987), it should be clear that slum dwellers' occupancy of hillsides is less voluntary than that, say, of the corporate executive who lives in Topanga Canyon 'for the view'.[7] The urban poor use their location as the base around which they organize livelihood activities (casual labour, street-trading, crafts, crime, prostitution). If the structure of urban landownership and rent means that the closest they can get to economic opportunities is a hillside slum, people will locate there regardless of the landslide risk (Hardoy and Satterthwaite 1989). This, we will argue, is a situation in which neither 'voluntary choice' models nor the notion of 'bounded rationality' is applicable (see Chapter 8).

Thirdly, the consequences of a mudslide for the rich are far less than for the surviving poor. Homes and possessions of the rich are usually insured (at least partly), while those of the poor tend not to be. The rich are more easily able to find alternative shelter and to continue with their income-earning activities after the disaster. They have reserves and credit available, and insurance.[8] The poor, by contrast, often have their entire stock of capital (home, clothing, tools for artisanal production, etc.) assembled at the site of the disaster. They have few if any cash reserves and are generally not considered creditworthy. Moreover, as emphasized above, location itself is a livelihood resource for the urban poor. In places where workers have to commute to work over distances similar to those habitually covered by the middle class, transport can absorb as much as 40 per cent of household budgets. The poor self-employed or casually-employed underclass finds such transport expenses insupportable. It is therefore not surprising that large numbers of working-class Mexicans affected by the 1985 earthquake refused to be relocated to the outskirts of Mexico City (Robinson *et al.* 1986).

ORIENTATION, SCOPE, AND PLAN OF THE BOOK

Most work on disasters emphasizes the importance of geo-tectonics, climatological, or biological 'trigger' events that arise in nature (recent examples include Bryant 1991; K. Smith 1992). Others focus on the human response (Dynes, De Marchi, and Pelanda 1987), psychosocial and physical trauma, economic, legal, and political consequences. Both these sets of literature assume that disasters are departures from 'normal' social functioning, and that recovery means a return to normal.

This book differs considerably from such conventional treatments of disaster, and arises from an alternative approach that has emerged in the last twenty years. This approach does not deny the significance of natural hazards as trigger events, but puts the main emphasis on the various ways in which social systems operate to generate disasters by making people vulnerable. In the 1970s and early 1980s, the vulnerability approach to disasters began with a rejection of the assumption that disasters are 'caused' in any simple way by external natural events, and a revision of the assumption that disasters are not 'normal'. Emel and Peet (1989) and Oliver-Smith (1986a) review these reactions. The competing vulnerability framework arose from the experience of research in situations where 'normal' daily life was itself difficult to distinguish from disaster. This work related to earlier notions of 'marginal' populations that emerged in studies of countries like Bangladesh, Nepal, Guatemala, Honduras, Peru, Chad, Mali, Upper Volta (Burkina Faso), Kenya, and Tanzania.[9]

Until the emergence of the idea of vulnerability to explain disasters, there was a range of prevailing views, none of which really dealt with the issue of how society creates the conditions in which people face hazards differently. One approach was unapologetically naturalist (sometimes termed physicalist), in which all blame was apportioned to 'the violent forces of nature' (Frazier 1979; Foster 1980). Other views of 'man [sic] and nature' (e.g. Burton, Kates, and White 1978; Whittow 1980; Maybury 1986) involved a more subtle environmental determinism, in which the limits of human rationality and consequent misperception of nature lead to tragic misjudgements in our interactions with it. 'Bounded rationality' was seen to lead the human animal again and again to rebuild its home on the ruins of settlements destroyed by flood, storm, landslide, and earthquake.

According to such views it was the pressure of population growth and lack of 'modernization' of the economy that drove human conquest of an unforgiving nature. This approach usually took a 'stages of economic growth' model for granted. Thus 'industrial' societies had typical patterns of loss from, and protection against, nature's extremes, while 'folk' (usually agrarian) societies had others, and 'mixed' societies showed characteristics in between (Burton, Kates, and White 1978). It was assumed that 'progress' and 'modernization' were taking place, and that 'folk' and 'mixed' societies would become 'industrial', and that we would all eventually enjoy the relatively secure life of 'postindustrial' society.

The 1970s saw increasing attempts to use 'political economy' to counter modernization theory and its triumphalist outlook, and 'political ecology' to combat increasingly subtle forms of environmental determinism.[10] These approaches also had serious flaws, though their analyses were moving in directions closer to our own than the conventional views.

We try to reintroduce the 'human factor' into disaster studies with greater precision, yet avoiding the dangers of an equally deterministic approach rooted in the political economy alone. We avoid notions of vulnerability that do no more than identify it with 'poverty' in general or some specific characteristic such as 'crowded conditions', 'unstable hillside agriculture', or 'traditional rain-fed farming technology'.[11] We also reject those definitions of vulnerability that focus exclusively on the ability of a system to cope with risk or loss.[12] These positions are an advance on environmental determinism, but lack an explanation of how one gets from very *widespread conditions* such as 'poverty' to very *particular vulnerabilities* that link the political economy to the actual hazards that people face. In Chapter 2 we begin our attempt to show these 'mechanisms' or dynamic processes that translate root causes into unsafe conditions.

The remainder of Chapter 1 and the next two chapters will set out the perspective of our book in detail. They describe how our view of disasters differs from the conventional wisdom, and also where they coincide. It is plainly wrong to ignore the role of hazards themselves in generating disasters, and the framework we are suggesting does not do so. Likewise, we are not suggesting that vulnerability is always the result of exploitation or inequality (just as it is not equivalent to poverty). It is integrally linked with the hazard events to which people are exposed. We also want to acknowledge that there are limits in this type of analysis. It is not always possible to know what the hazards affecting a group of people might be, and public awareness of long-return period hazards may be lacking. For instance, Mount Pinatubo in the Philippines erupted in 1991, but had been dormant for 450 years.

Chapter 2 introduces our alternative view of disasters. It uses a simple model of the way in which 'underlying factors' and root causes embedded in everyday life give rise to 'dynamic pressures' affecting particular groups, leading to specifically 'unsafe conditions'. Being at risk of disaster is shown to be the chance that the characteristics of people generated by these political–economic conditions coincide in time and space with an extreme 'trigger event' natural hazard to which they have been made vulnerable. This will be referred to as the 'pressure and release' model (PAR), since at first it is used to show the pressure from both hazard and unsafe conditions that leads to disaster, and then how changes in vulnerability can release people from being at risk.[13]

12

We consider that certain characteristics of groups and individuals have a great deal to do with determining their vulnerability to hazards. Some of these, such as socio-economic class, ethnicity, and caste membership have featured in analyses since the 1970s. Others, especially gender and age, are more recent as research categories, and have developed partly because of the influence of social movements such as feminism.[14] For example, Vaughan (1987: 119 47) uses the oral evidence provided in women's songs and stories in Malawi to reconstruct a women's history of the 1949 famine that is strikingly different from men's accounts:

> [Women], along with the very old and very young, were more likely than men to end up relying on government handouts . . . . Women stress how frequently they were abandoned by men, how harrowing it was to be left responsible for their suffering and dying children, how they became sterile, and how they were humiliated by the feeding system.
>
> (ibid.: 123)

Others have emphasized the special needs, lack of status and access, hence special vulnerability of the frail elderly, especially widows (Guillette 1991, 1992; Feierman 1985; Wilson and Ramphele 1989: 170–85).

The activities of daily life comprise a set of points in space and time where physical hazards, social relations, and individual choice converge.[15] Patterns of vulnerability emerge at this convergence, at which point several socio-economic and personal characteristics of people have a bearing on vulnerability to disaster. Here are found sometimes (but not always) the effects of gender,[16] age,[17] physical disability,[18] religion,[19] caste,[20] or ethnicity.[21] All of these may play a role in addition to poverty, class, or socio-economic status. Although we include class in our analysis, a major difference between our work and that of the earlier 'political economy/political ecology' school is that we fully recognize the role of this wide range of social relations and do not dwell exclusively on class relations.

Chapter 3 adds to our alternative framework by focusing on patterns of access to livelihood resources. We expand the discussion there of 'underlying factors and root causes' identified in Chapter 2. In doing so we seek to shift the analytical balance between society, population, and environment further in the social direction, without oversimplifying or producing a theory that is of little use to managers, planners, and policy-makers. There is much here of direct significance to policy-makers in an emphasis on the ways in which vulnerable people cope with patterns of access they face.

Part I concludes with a discussion of coping. We believe that too little attention has been given to the strategies and actions of vulnerable people themselves. In large part their 'normal' life is evidently (at least to outsiders) a continual struggle in which their conditions may resemble a disaster. People become braced to cope with extreme natural events through the stress of making ends meet, in avoiding the daily hazards of work and home,

13

and of evading the predations of the more powerful. They form support networks, develop multiple sources of livelihood access, and 'resist' official encroachments on livelihood systems in a variety of ways (J. C. Scott 1985, 1990). People learn rather cynically, yet realistically, not to rely on services provided by authorities (Robinson *et al.* 1986; O'Riordon 1986; Maskrey 1989). Our discussion of 'coping' will neither romanticize the self-protective behaviour of ordinary people, nor dismiss it.[22]

Having set out our alternative framework in Part I (Chapters 1 to 3), Part II presents case material organized by conventional hazard type – those linked with famine, pandemic disease and biological disasters, flood, cyclone, earthquake, volcano, and landslide (Chapters 4 to 8). In each chapter we follow a similar method in tracing the causes of vulnerability, making use of both 'pressure' (PAR) and 'access' frameworks. Although it may appear to contradict our approach to deal with disasters through different natural hazard types, we have deliberately chosen to do this because users of this book may themselves be concerned with particular hazards, or may find it difficult to accept our approach without seeing it interpreted more concretely in the context of nature.

Part III (Chapters 9 and 10) draws out lessons for recovery and reconstruction and for preventive action. We try to leave the reader with practical guidelines that indicate how 'vulnerability analysis' can be incorporated into routine precautionary and development planning.

LIMITS AND ASSUMPTIONS

Limitation of scale

There are logical grounds for limiting our book to certain sorts of disaster. Disasters cannot, of course, be neatly categorized either in type or scale. At one extreme, it seems that there have been five mass extinctions over the last 400 million years in which up to half of the life forms on the planet disappeared (Wilson 1989: 111). The best known of these is the disappearance of the dinosaurs. The scale of such disasters (and even the use of the term is perhaps inappropriate) is clearly so many orders of magnitude greater than those with which we are concerned that we exclude them. Such events are beyond the present scale of human systems, and are outside the processes which interest us as ones which humankind can influence if it wishes.[23]

More recently, there have been two or three occasions when a large proportion of the human inhabitants of this planet died with apparently little distinction in regard to the relative risk of different social groups. Many millions died during the pandemics of bubonic and pneumonic plague known as the Plague of Justinian (AD 541–93) and the Black Death (1348–53). More recently, the influenza virus that swept the world during the First World War killed 22 million in less than two years (1918–19). This was approximately four times the total of military casualties during that war. The demographic and socio-economic consequences of the first two events had epochal significance. The current AIDS pandemic could equal them in its widespread socio-economic consequences unless a vaccine is found or sexual practices change.

Despite the great significance of biological disasters, we shall address such events only tangentially (Chapter 5) and to the limits of usefulness of the vulnerability approach, treating them more as limiting cases that shed light on 'normal' disasters, such as outbreaks of cholera in Latin America and Africa.

Nuclear war is another type of disaster which we do not consider because it is so obviously man-made, although research on 'nuclear winter' has been inspired by threats from natural events such as massive volcanic explosions or asteroid impacts. There is also considerable climatological, astrophysical, and palaeontological work on mass extinctions which links some to severe interference with received solar radiation. On the other hand, atmospheric phenomena of a similar magnitude, such as global warming, will be treated as part of the more remote 'dynamic pressures' of the PAR model, shaping patterns of vulnerability. We also consider war (in its 'normal', non-nuclear form) itself to be a significant 'root cause' of disaster and will address it at many places in the text.

Technology and human hazards

Vulnerability analysis may be relevant to analysing disasters resulting from man-made hazards. But we restrict the scope of the book to exclude technological hazards, for the simple reason that they are obviously not natural in origin. One of our purposes in this book is to deal with natural hazards because of the inadequacy of explanations of disasters which blame nature. Our aim is to demonstrate the social processes that, through people's vulnerability, generate human causation of disasters from natural hazards. So there is little point in looking at specifically human-created hazards.

Failure of nuclear technology such as occurred at Chernobyl (or, very nearly, Three Mile Island in the USA, or Windscale in Britain), or massive oil and toxic spills, are excluded on the grounds of falling outside this focus. Such technological hazards are discussed by a number of others (Zeigler, Johnson, and Brunn 1983; Perrow 1984; Kirby 1990a; Button 1992). Later, there will be some discussion of the Bhopal disaster in India, which involved explosions at a toxic chemical factory. The same locational factors responsible for generating hillside slums already mentioned in other countries led to a dense squatter settlement around the plant. Such a case is at the limits of our type of analysis, and overlaps with a related literature concerning technology and society (Weir 1987, Davis 1984b).

15

What happens to poor and other vulnerable people who find themselves in the path of rapid industrialization, deindustrialization, industrial deregulation, or importation of toxic waste is clearly of concern to us, but cannot be the central issue in this book. However, some overlap with a critical appraisal of technological risk and so-called 'modernization' will nevertheless occur in the chapters that follow. Flooding caused by the failure of a dam is a good example (Chapter 6). The web of cause and effect in the connections between society, nature, and technology is often impossible to disentangle.

We will be concerned with the impact of technology on vulnerability, particularly technology in its apparently simplest and most benign forms.[24] For example, a new road may link a previously isolated rural community with sources of food which may reduce vulnerability in times of drought. That same road may also lead away able-bodied youth in search of urban income, reducing the labour available to maintain traditional earth and stone works constructed to prevent erosion, or to build or repair houses adequately to withstand earthquake. The result may be reduction of crop yield during drought years because of additional soil loss or deaths from earthquake which otherwise would be preventable. The same road may introduce mobile clinics that immunize children against life-threatening diseases, or it may provide the channel through which 'urban' diseases such as tuberculosis and sexually-transmitted diseases arrive with men who had gone to work in city, mine, or plantation.

The very same technological artifact, the road, may provoke landslides that kill people or reduce the available arable land. All these contradictory effects of technological change are possible. The same may be said of the introduction of new water or energy sources or new seed varieties. There are several ways such questions of technological change arise in relation to disaster vulnerability.

One of the most frequent responses to disaster by outsiders is the provision of various technologies to the affected site during relief and rehabilitation activities. These include temporary housing, food supplies, alternative water supplies and sanitation facilities, seeds, and tools to re-establish economic activities. In all such cases, the new or temporary technology may have a role in increasing or decreasing the vulnerability of some social group to a future hazard event.

Even more generally, development planners often introduce technology at the so-called 'leading edge' of whatever version of rapid, systemic change they define as 'development'. This may be irrigation technology in the form of a large dam that displaces thousands of families in what economists call 'the short run'. It might take the form of low-income housing or the development of an industrial complex. Such development initiatives can have a series of unintended, unforeseen consequences.

The people displaced following the flooding behind a large dam may not benefit from resettlement in the areas that are fed by the irrigation water, or if they are included among settlers, they may end up at the bottom of the water distribution system where water is scarce. Women on such new schemes may lose conventional rights to land on which they used to grow food for their families (Rogers 1980) or their knowledge and skills may be rendered 'obsolete' (Shiva 1989). Nutritional levels among children may fall, paradoxically, as cash income from the marketed product of irrigation increases (Bryceson 1989).

The introduction of technology can modify and shift patterns of vulnerability to hazards. For example, the Green Revolution varieties of grain have shifted the risk of drought and flood from an emergent class of 'modern' farmers to the increasing number of landless and land-poor peasants. The latter have become more vulnerable because they are denied access to 'commons' that formerly provided livelihood resources and because they are highly dependent on wages earned in farm labour to purchase food and other necessities (Jodha 1991; Chambers, Saxena, and Shah 1990; Shiva 1991). They are also vulnerable because they now depend for food and other basic necessities on wages from farm employment that can be interrupted by flood, hail, drought, or outbreak of pests (Drèze and Sen 1989; see Chapter 4).

The increased output of grain due to the Green Revolution in many parts of South Asia may have radically reduced the threat of famine for many (mostly urban) people, but it has also contributed to malnutrition and increased vulnerability for the resource-poor in rural areas. The change in technology has affected a pre-existing social and economic structure that has not been able to distribute any benefits properly, and has led to a realignment of assets and income. The losers may be subject to new hazards as a result. For example, they may migrate into low-lying coastal land exposed to storms in order to find land (see Chapter 7). The literature on development is full of studies of such unintended consequences.[25]

This book will focus on such technological developments and their consequences where they can be seen to impinge on people's vulnerability to extremes in nature, or where they affect the ability of groups to sustain livelihood in the aftermath of environmental extremes. We will not, therefore, be concerned with the risk created directly by the new technology such as the actual mechanical failure of the dam or the catastrophic failure of the factory (such as the one in Bhopal), but rather with the ways in which that technology modifies livelihoods and their sustainability.

NOTES

1 This comprises the International Federation of Red Cross and Red Crescent Societies (IFRCRCS), with its world headquarters in Geneva and a widespread national and sub-national system of training in public health, safety, and emergency response.

2 Recent self-critical evaluations by voluntary agencies include one by a broad coalition that supported 'Operation Lifeline Sudan' (Minear 1991) and the group 'USA for Africa' (Scott and Mpanya 1991).

3 On non-governmental organizations (private voluntary organizations, popular development organizations, development support organizations, etc.) see Conroy and Litvinoff (1988); Holloway (1989); During (1989); J. Clark (1991).

4 This is the layer of the earth's mantle upon which the lithospheric plates sit. Convection currents in the asthenosphere allow heated material to rise, while cool material sinks, leading to movement of the plates. Understanding of biogeo-chemical cycling and plate tectonics (including earthquakes and volcanoes) would require study of the asthenosphere as well as the more accessible lithosphere.

5 The World Commission on Environment and Development (the Brundtland Commission) linked the concept of livelihood to the ability of people to protect the environment, and stated that the goal of development should be 'sustainable livelihood security' (WCED 1987a; cf. Chambers 1983). In our view, vulner-ability to hazards is likely to increase when livelihoods are pursued at the expense of environmental stability. So it is not a solution to vulnerability if people seek to increase their access to livelihood resources for short-term gains, even if it is necessary to cope with the immediate impact of hazards.

6 In 1991 and 1992 there were torrential rains and mudslides in southern California affecting two counties (Ventura and Los Angeles) where 10 million people live. 1991 also saw a fire storm that killed twenty-five people and left thousands homeless in the middle income, suburban hills above Oakland and Berkeley in northern California. During this same period there were a number of mudslides in Rio de Janeiro and Belo Horizonte in the industrial south of Brazil (see Chapter 8).

7 In Chapter 8 a similar situation is discussed. There is a very dangerous active volcano in the middle of Lake Taal in the Philippines. Despite its known history of recent eruptions, many rich Filipinos, including former President Marcos, had built luxury second homes on the shores of lake Taal 'because of the view'. This behaviour differs from that of the inhabitants of the volcanic island – also very much at risk – who are there to make a living. In 1992 the island was partially evacuated when the volcano showed signs of erupting.

8 In 1991 thousands of middle-class people's homes were destroyed by wildfire in the Oakland/Berkeley hills. They grouped together in three Phoenix Associations that have enough economic power to negotiate special prices for wholesale building materials, and enough political clout to get the city council to block one developer (who wanted to rebuild with high-density housing) by declaring the area a park.

9 Later, in Chapter 9, we suggest that a major watershed for relief agencies was the year 1970, when enormous disasters in Peru, East Pakistan (now Bangladesh), and Biafra (Nigeria) coincided. Subsequent reflections on these events plus the Sahel famine (1967–73) and drought elsewhere in Africa, erosion in Nepal, earthquake in Guatemala (1976), and hurricane affecting Honduras (1976) led to several attempts at synthesis and a new 'theory' of disasters that focused on the vulnerability of 'marginal' groups (Meillasoux 1973; Baird et al. 1975; Blaikie, Cameron, and Seddon 1977; Davis 1978).

10 On the response of 'political economy' and 'political ecology' to both 'moderni-zation theory' and 'environmental determinism' see Meillasoux (1973); Baird et al. (1976); Wisner, O'Keefe, and Westgate (1977); Susman, O'Keefe, and Wisner (1983). Work during this period was heavily influenced by Latin American dependency theory.

11 For examples of the use of a too general notion of vulnerability, see Anderson and Woodrow (1989); Parry and Carter (1987); Cuny (1983); Davis (1978). In such cases it is essential to specify the mechanisms by which one gets from generally widespread conditions (e.g. 'poverty' or 'crowded conditions') to particular vulnerabilities (e.g. to mudslide, cyclone, earthquake, famine).

12 Such functionalist views of social system coping include those of Mileti, Drabek, and Haas (1975); Timmerman (1981); Pelanda (1981); Drabek (1986); and J. Lewis (1987). On the whole we take the view that one has to be more specific. *People* cope, not disembodied systems. See Chapter 3.

13 This view has much in common with other recent attempts to reconcile an analysis of structural constraints on people's lives with an appreciation of the individual's agency and freedom (Mitchell 1990; Palm 1990; Kirby 1990b).

14 The women's movement makes an enormous contribution to our understanding of vulnerability, environmental degradation, and the possibilities for restoration, peace-making, and 'healing'. This often requires redefining what is meant by such terms as 'development' and 'progress'. See Sen and Grown (1987); Momsen and Townsend (1987); Dankelman and Davidson (1988); Shiva (1989); Tinker (1990); and Cliff (1991), on women and the politics of 'development' and vulnerability, as well as eco-feminist philosophers Merchant (1989); and Biehl (1991).

15 Accounts of disaster that try to balance macro- and micro-perspectives include Hewitt (1983a); Oliver-Smith (1986b); G. Kent (1987); Maskrey (1989); Kirby (1990b, 1990c); Palm (1990).

16 Studies emphasizing the role of gender in structuring vulnerability include Jiggins (1986); Schroeder (1987); M. Ali (1987); Rivers (1982); Vaughan (1987); Drèze and Sen (1989: 55–9); Sen (1988, 1990); Agarwal (1990); Kerner and Cook (1991); O'Brien and Gruenbaum (1991).

17 The very young are highly vulnerable to nutritional and other health stress during disasters (Chen 1973; UNICEF 1989; Goodfield 1991). The old are often more vulnerable to extremes of heat and cold and are less mobile and capable of evacuation (O'Riordon 1986: 281; Bell, Kara, and Batterson 1978), and are particularly vulnerable to recurrent disasters (Guillette 1991). Widows in many parts of the world are especially vulnerable, as in southern Africa (Wilson and Ramphele 1989: 177–8; Murray 1981) and East Africa (Feierman 1985).

18 Disabilities such as blindness, mental retardation, somatic hereditary defects, and post-traumatic injury (such as spinal cord injuries) affect hundreds of millions of people worldwide (Noble 1981). People with disabilities have specific increased vulnerabilities in the face of hazards due to their impaired mobility, or interruption of the special attention to their hygiene and continuous health care needs in disasters (UNDRO 1982a; Parr 1987).

19 The role of religion has not been as well studied, but consider recent events. The Burmese fleeing into Bangladesh during 1992 were a Muslim minority in their home country. The 400,000 people forced to leave squatter settlements around the city of Khartoum for an uncertain future in 'resettlement camps' in the desert were mostly a Christian or animist minority, refugees from war in the south, in the predominantly Muslim north of Sudan.

20 The role of caste has been most fully explored in studies of famine in India (see Chapter 4). However, there is also a suggestion that caste-based locational segregation homes in rural and urban India may have a bearing on vulnerability to riverine flood and cyclone (Chapters 6 and 7).

21 Ethnicity emerges as an important factor in explaining vulnerability in studies by Regan (1983); Franke (1984); Perry and Mushkatel (1986); Winchester (1986, 1990); Laird (1992); Miller and Simile (1992); Johnston and Schulte (1992).

22 Perceptions of risk are sometimes deeply-rooted in cultural understandings of ritual purity and danger (Douglas and Wildavsky 1982); complaints about authorities and claims of suffering can sometimes be gambits in complex games over local political power (Richards 1983; Laird 1992).

23 In fact, a NASA team reported to the US Congress in 1992 on the probability and consequences of an asteroid colliding with the earth. They put the risk at one collision every 300,000 to 1 million years and warned that asteroids with diameter of 1 km (0.62 miles) and larger could scatter enough pulverised rock and dust to block most of the sun's light and destroy agriculture. Apparently such asteroids are not rare, and the earth missed hitting one by 6 hours (their paths came that close) in 1990 (Broad 1992)

24 At the other extreme from 'simplest and most benign' we *do* treat risks associated with biotechnology in Chapter 5. Also, it is hard to disentangle risks associated with construction technologies (Chapter 8) or agricultural innovations (Chapter 4) from such hazards as earthquake and famine.

25 The unintended consequences of 'development' are documented by Trainer (1989); Shiva (1989); Wisner (1988b); Lipton and Longhurst (1989). Special note should be taken of a 'classic' early paper on disease and development by Hughes and Hunter (1970) and the contrast with the role of other kinds of 'development' in restoring health of communities (Wisner 1976a).

2

DISASTER PRESSURE AND RELEASE MODEL

THE NATURE OF VULNERABILITY

Two models

In evaluating disaster risk, the social production of vulnerability needs to be considered with at least the same degree of importance that is devoted to understanding and addressing natural hazards. Expressed schematically, our view is that the risk faced by people must be considered as a complex combination of vulnerability and hazard. Disasters are a result of the interaction of both; there is no risk if there are hazards but vulnerability is nil, or if there is a vulnerable population but no hazard event.

'Hazard' refers to the extreme natural events which may affect different places singly or in combination (coastlines, hillsides, earthquake faults, savannas, rain forests, etc.) at different times (season of the year, time of day, over varying return periods, of different duration). The hazard has varying degrees of intensity and severity. Although our knowledge of physical causal mechanisms is incomplete, some long records (for example, of hurricanes, earthquakes, snow avalanches, or droughts) allow us to specify the statistical likelihood of many hazards in time and space. But such knowledge is of very limited use in calculating the actual level of risk. What we are arguing is that risk is a compound function of this complex (but knowable) natural hazard and the number of people characterized by their varying degrees of vulnerability who occupy the space and time of exposure to extreme events.

A disaster occurs when a significant number of vulnerable people experience a hazard and suffer severe damage and/or disruption of their livelihood system in such a way that recovery is unlikely without external aid. By 'recovery' we mean the psychological and physical recovery of the victims, the replacement of physical resources and the social relations required to use them.

In order to understand risk in terms of our vulnerability analysis in specific hazard situations, this book uses two related models of disaster. The 'pressure and release model' (PAR model) is introduced in this chapter as a relatively simple tool for showing how disasters occur when natural hazards

21

affect vulnerable people. Their vulnerability is rooted in social processes and underlying causes which may ultimately be quite remote from the disaster event itself. It is a means for understanding and explaining the causes of disaster.

The basis for the pressure and release (PAR) idea is that a disaster is the intersection of two opposing forces: those processes generating vulnerability on one side, and physical exposure to a hazard on the other. The image resembles a nutcracker, with increasing pressure on people arising from either side – from their vulnerability and from the impact (and severity) of the hazard on those people at different degrees of vulnerability. The 'release' idea is incorporated to conceptualize the reduction of disaster: to relieve the pressure, vulnerability has to be reduced. In fact this chapter focuses on the pressure aspect of the PAR model, and the discussion of conditions for creating release is left mainly for Part III.

A second model, referred to as the 'access model' is discussed in Chapter 3. In effect it is an expanded analysis of the principal factors in the PAR model that relate to human vulnerability and exposure to physical hazard. It is a more magnified analysis of how vulnerability is generated by economic and political processes. It indicates more specifically how conditions need to change to reduce vulnerability and thereby improve protection and capacity for recovery. It also avoids the oversimplification of the PAR model, which suggests (in its image of two separate sides in the diagram) that the hazard event is isolated and distinct from the conditions that create vulnerability.

As will be seen in the access model, hazards themselves do of course also alter the set of resources available to households (e.g. through flood destruction of crop or land) and alter the patterns of recoverability of different groups of people. Hazards actually intensify some people's vulnerability, and the incorporation of this insight provides a significant improvement on ideas that see disasters simply as the result of natural events detached from social systems.

CAUSE AND EFFECT
IN THE DISASTER PRESSURE MODEL

The following section anticipates Part II, where the chain of explanation of disasters will be related to a series of different types of hazard.

The chain of explanation

Figure 2.1 illustrates the pressure and release model, and is based on the idea that an explanation of disaster requires us to trace a progression that connects the impact of a hazard on people through a series of levels of social factors that generate vulnerability.[1] The explanation of vulnerability has three such links or levels which connect the disaster to processes that are sometimes quite remote and lie in the economic and political sphere.

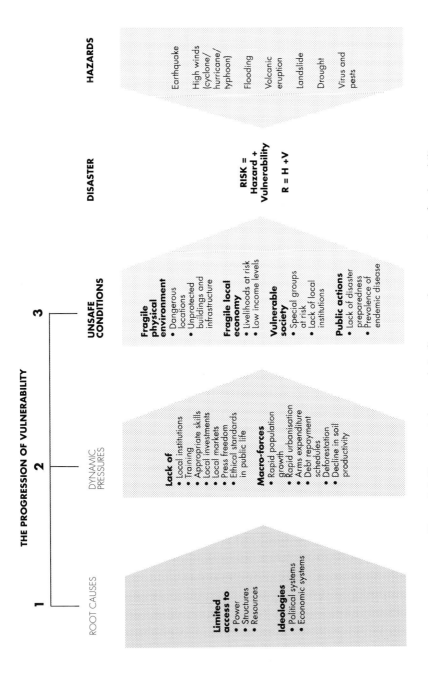

Figure 2.1 'Pressures' that result in disasters: the progression of vulnerability

The most distant of these are *root causes* (or underlying causes), a set of well-established, widespread processes within a society and the world economy. The most important root causes that give rise to vulnerability (and that reproduce vulnerability over time) are economic, demographic, and political processes. These affect the allocation and distribution of resources between different groups of people.

Such root causes are normally a function of economic structure, legal definitions of rights, gender relations, and other elements of the ideological order. They are connected with the functioning (or not) of the state, and ultimately control by police and military. Military force sometimes has its own impact as an underlying cause of disasters such as famine, especially in prolonged so-called low-intensity warfare (Clay and Holcomb 1985; E. Hansen 1987). Recent examples are all too common and include Somalia, Sudan, Ethiopia, Chad, Liberia, Angola, and Mozambique, where long drawn out wars and famine coincide. We will return to war as a factor in disaster vulnerability at the end of this chapter.

Root causes reflect the distribution of power in a society. People who are economically marginal (such as urban squatters) or who live in 'marginal' environments (isolated, arid or semi-arid, coastal, or forest ecosystems) tend also to be of marginal importance to those who hold economic and political power (Blaikie and Brookfield 1987: 21–3; Wisner 1976b, 1978b, 1980). This creates two sources of vulnerability for such groups. Firstly, their access to livelihoods and resources that are less secure and rewarding is likely to generate higher levels of vulnerability. Secondly, they are likely to be a low priority for government interventions intended to deal with hazard mitigation.

Dynamic pressures are processes and activities that 'translate' the effects of root causes into the vulnerability of unsafe conditions. Dynamic pressures channel the root causes into particular forms of insecurity that have to be considered in relation to the types of hazards facing those people. These include reduced access to resources as a result of the way regional or global pressures such as rapid population growth, epidemic disease, rapid urbanization, war, foreign debt and structural adjustment, export promotion, mining, hydropower development, and deforestation work through to localities.

The way these dynamic pressures operate to channel root causes into unsafe conditions can be seen clearly with the example of endemic disease and malnutrition. People's basic health and nutritional status relates strongly to their ability to survive disruptions of their livelihood system. This status is important for their 'resilience' in the face of external shock. Chronically undernourished and diseased populations succumb sooner in famine than do previously well-nourished and healthy ones. The relation between nutrition and disease is such that chronically malnourished people have less active immune systems and suffer more from infections common in disasters such as measles or dysentery. The age distribution of a population is also of importance: children and the frail elderly suffer more from certain stresses (hunger, extreme heat and cold) during disruptions of their livelihood system.

Rural–urban migration is another dynamic pressure which arises in many parts of the Third World in response to the economic framework inherent in root causes. This may lead to the erosion of local knowledge and institutions required for coping in the aftermath of a disaster. The loss of younger people, especially working-age males and those with skills which are marketable in the cities (or even abroad), may alter the type of building structure that can be constructed to something less safe than previously. Later in this chapter in the section on global dynamic factors we discuss in more detail how a number of root causes (such as population growth, urbanization, economic pressures, foreign debt, land degradation, global environmental change, and war) are 'channelled' or 'translated' by dynamic pressures into particular unsafe conditions.

Unsafe conditions are the specific forms in which the vulnerability of a population is expressed in time and space in conjunction with a hazard. Examples include people having to live in dangerous locations, being unable to afford safe buildings, lacking effective protection by the state (for instance, in terms of effective building codes), having to engage in dangerous livelihoods (such as ocean fishing in small boats, or wildlife poaching, or prostitution, with its health risks), or having minimal food entitlements, or entitlements that are prone to rapid disruption.

We propose the following terminology when dealing with unsafe conditions. People, as should be apparent already, are *vulnerable* and live in or work under unsafe conditions ('unsafe' can refer to locations of work or habitation, where people spend their daily lives). We avoid using the word vulnerable in regard to livelihoods, buildings, settlement locations, or infrastructure, and instead use terms like 'unsafe', 'fragile', 'unstable' or their synonyms.

The chain of explanation linking unsafe conditions to dynamic pressures and root causes can be illustrated by looking at some ideas of unsafe locations. The presence of people in hazardous places is often the result of broader political economic pressures, which may displace weaker groups. Chapter 8, on earthquake, provides examples of this. Similar situations are found in relation to drought in Africa. Some of the West African nomadic Tuareg now inhabit a zone of very high rainfall variability. Some were displaced fifty years ago by the expansion of groundnut (peanut) cultivation encouraged by the French colonial state (Franke and Chasin 1980).

In Kenya there are more than 2 million people trying to farm on the margins of rangelands where rainfall is low (averaging less than 500 mm (20 inches) per year) and highly variable. They were displaced from highlands in the centre and west of the country during the post-colonial

Africanization of settler farms in the 1960s and 1970s (Wisner 1976b, 1978b, 1980). In Indonesia there are thousands of poor farm families in the tropical forests of Kalimantan who both caused and suffered massive forest fires in the late 1980s. They were the displaced victims of the operation of a land market elsewhere in Indonesia (especially Java and Bali). Rather than drift to the cities, they opted for official 'transmigration' schemes or simply followed roads built by lumberjacks penetrating the forest.

In each of these cases the unsafe location is linked by a series of dynamic pressures that can be traced back to root causes. This is illustrated in Figure 2.1, where vulnerability that arises from unsafe conditions intersects with a physical hazard (trigger event) to create a disaster, but is itself only explained by analysis of the dynamic processes and root causes which generate the unsafe conditions. It is important to note that by 'cause and effect' we do not imply that single causes give rise to single effects. In their study of land degradation, Blaikie and Brookfield (1987) refer to such causal sequences as 'cascades'. There are many ways in which dynamic processes (some unique to particular societies, some nearly universal because of the pervasive influence of global forces) channel root causes into unsafe conditions and to specific time–space collisions with a natural hazard. This can be illustrated in the outcome of floods in Bangladesh in 1988 (see Box 2.1), and landslide and earthquake impacts in part of north Pakistan (Box 2.2).

TIME AND THE CHAIN OF EXPLANATION

Root causes, dynamic pressures, and unsafe conditions are all subject to change, and in many cases the processes involved are probably changing faster than in the past. The change in building techniques and material in Pakistan was rapid, as were the processes of outmigration and deforestation, and affected communities that had changed little for many years. Even large-scale processes, such as population growth, are rapid by comparison with changes in, say, values and beliefs or in legal structures. An annual population increase of 4.2 per cent as in Kenya, giving a doubling time of sixteen years, must be considered a significant factor channelling the root causes of vulnerability into unsafe conditions.

Location and livelihood can change even more rapidly. For example, between 1973 and 1976 about half of the then 12 million rural inhabitants of Tanzania were variously encouraged or coerced into nucleated villages (Coulson 1982). This completely altered settlement patterns and the resource basis of the affected people's livelihoods over a period of only three years. Other forms of such disruptions are common as a result of war. During the peak of wars and famines in Sub-Saharan Africa in 1983–5, 5 million refugees of one kind or another were forced to live outside their national boundaries (CIMADE, INODEP, and MINK 1986), and uncounted others remained within them. The impact of these disruptions of

26

Box 2.1   Landless squatters in Dhaka

Dhaka, the capital of Bangladesh, is situated in the flood plain of a major river, the Buriganga. To the northwest is a large zone of low-lying, flood-prone land in the vicinity of Nagor Konda. Here squatter settlements have grown rapidly, as they have in many areas around the capital in recent years (Shaker 1987). It has been densely settled, particularly since 1970, mostly by poor landless families from the south and east of the country (Rashid 1977). The former landless who now inhabit this depression do so because of proximity to Dhaka vegetable market. Already the chain of explanation of their vulnerability can be seen at work: rural landlessness channels people who have few alternatives to try seeking a particular economic opportunity provided by the urban vegetable market. But this gives rise to their occupation of an unsafe location. As relative newcomers, and very poor people, the squatters in these low-lying areas do not have access to the structures of power that control marketing. They also have insecure title to land in the depression, so they are unable to borrow in order to increase their productivity and compete with better-established market gardeners (M. Ali, pers. comm.; cf. M. Ali 1987).

This situation means that they have to grow rice more than vegetables in the land of the depression. For although it is near the city, the poor squatting at Nagor Konda cannot compete with vegetable-growers who have better access to credit and to markets, and are thus forced into low-income pursuits. They are unlikely to make any savings, and grow rice for subsistence, marketing only a small amount of vegetables.

On the eve of Bangladesh's massive floods in August 1988, this relatively powerless group was living in an economically marginal situation but close to the city, on low-lying land prone to flooding. Their economic and political marginality meant that they had few assets in reserve. It also meant that their children were usually malnourished and chronically ill. This channelled the dynamic pressures arising out of landlessness and economic marginalization into a particular form of vulnerability: lack of resistance to diarrhoeal disease and hunger following the flooding in 1988. Factors involving power, access, location, livelihood, and biology mutually determined a situation of particular unsafe conditions and enhanced vulnerability. These social, economic, and political causes constitute one side of the pressure model. The other, the floods themselves during August 1988, constitute the trigger event whose impact on vulnerable people created the disaster.

---

access on the vulnerability of these peoples to drought and other hazards is hardly studied.

The time factor is not only influenced by accelerating rates of change affecting livelihood systems and generating vulnerability. Global factors involved in our 'root causes' shift and alter at varying rates, and interact with one another in complex ways, the outcome of which is unpredictable. One set of root causes may lead to dynamic pressures of different types at different times and places. Such variations in the dynamic processes that generate vulnerability may relate more strongly to one sort of hazard and not another.

Box 2.2 Karakoram hazards

This case comes from an interdisciplinary study of the housing safety in the Karakoram area of northern Pakistan (Davis 1984a; D'Souza 1984). In this we follow the chain of explanation that links vulnerability to the specific physical trigger that creates a disaster in reverse, starting with 'unsafe conditions'.

The research team carefully examined local dwellings and settlement patterns within the context of a rural economy. They found that the communities were at risk from a wide range of hazards. In this region traditional dwellings were built with stone masonry walls. A series of timber bands were set at regular intervals in the height of each wall, in order to hold the stones together, and the complex timber roofs were constructed with a very heavy covering of earth to provide much-needed insulation.

These traditional dwellings were built until around the 1960s or early 1970s, and provided some protection against earthquakes. But subsequently local building patterns changed in favour of concrete construction. The new houses were intended to be reinforced, but in reality they were built without any real understanding of how to connect steel to concrete or roofs to walls. The siting of most buildings was equally dangerous, since to avoid reducing their meagre landholdings (all available flat land was used for agriculture), many houses were built on exceedingly steep slopes posing risks from landslides and rockfalls.

The net result was a highly dangerous situation, and the failure to provide protection was due to a number of interlocking influences which together produced these unsafe conditions. These factors included reduced concern about building safety due to other issues or 'risks' which were of a more everyday nature taking priority, so that less money was spent on dwellings. There was also a lack of knowledge of both concrete construction and aseismic (shock-proof) techniques, a shortage of skills, and a change in the availability of building materials.

In turn, some of these (especially the lack of availability of both skills and materials) could be directly attributed to 'dynamic pressures'. Firstly, a shortage of timber for building and other purposes had arisen in the region, related to the effects of the rising population in the area. This had led to a rapid increase in tree-cutting for firewood in the cold climate, and to create additional fields for cultivation. There were outside pressures on forests as well, from commercial logging operations.

Secondly, there was a very serious shortage of skilled carpenters and masons, so buildings were constructed and maintained by farmers and labourers who freely admitted that they knew very little about the task. In trying to piece together the reasons for the absence of good builders another dynamic pressure emerged. During the 1970s the Chinese government had built the Karakoram Highway, a major access road into the area. This linked China with the Pakistani capital, Islamabad. The road was built for political and strategic reasons, but it was intended to bring 'development' to the remote northern areas. Risk was 'imported' via the highway to the extent that heavy (unsafe) concrete buildings were developed and considered 'modern' (Coburn et al. 1984). There was also a migration of carpenters out of the area, by means of this road, to Karachi, Islamabad, and even to the Gulf region

Box 2.2 continued

(where earnings were twenty times higher). As so often happens, while the road was being used to bring in medical and educational resources, it also enabled loggers to enter the region for the first time, and they removed vast quantities of timber, often by theft. It is likely that the resulting deforestation has contributed to soil erosion and slope instability.

The reason the Chinese provided aid for the Karakoram Highway is rooted in regional conflicts and China's alliance with Pakistan and hitherto suspicion of India, issues themselves related to Cold War rivalries. The Pakistan government encouraged their much-needed workforce to migrate from the country so as to attract the foreign currency remittances sent for family support by the workers abroad. This would reduce the country's balance of payments deficit.[2] In this way we are led from proximate and specific cause to more remote 'root causes'. The net result was that these families were left to live in dangerous homes, in villages depleted of the men who otherwise would have maintained them and improved their families' safety.

Root causes often shift because of disputed power, and vulnerability may therefore change as a result. The converse is also true. Mass suffering due to disaster may contribute to the overthrow of elites, and lead to dramatic realignments of power. It can be argued that the cyclone and storm surge in East Pakistan in 1970 contributed to the development of the Bangladesh independence movement, that governments in Niger and Ethiopia were overthrown partly as a result of their behaviour in the 1970s Sahel famine, and that the revolutionary movement in Nicaragua from 1974–9 got some of its impetus from the effects of the Managua earthquake of 1972.

LIMITS TO OUR KNOWLEDGE

Vulnerability can be defined reasonably precisely. 'Unsafe conditions' have been the subject of much detailed research, and in much of the world detailed knowledge has been obtained about which sites might fail in a landslide, or which buildings will survive or collapse in an earthquake and why. Similarly, the processes involved in dynamic pressures and root causes are reasonably well understood in many situations, as, for instance, for urbanization and technological change.

However, as we move back from unsafe conditions and vulnerability to root causes, the directness of the linkages (and therefore the level of precision in disaster explanation) becomes less definite. In analysing the linkages between root causes, dynamic pressures, and unsafe conditions, it is rather exceptional to have reliable evidence, especially the further back in the chain of explanation we go.

These gaps in knowledge concerning the linkages of underlying causes or pressures on vulnerability are serious. This partly explains why unsafe conditions have arisen and are allowed to persist. At best this lack of understanding is likely to result in policy-makers and decision-takers, restricted by the scarce resources at their disposal, addressing incorrect pressures or causes. At worst, it provides additional excuses for apathy or a continuation of cosmetic approaches.

Yet to some extent these gaps exist because of the failure to ask the right sort of questions. It is imperative to accept that vulnerability involves something very different from simply dealing with hazards through mitigation, prediction, or relief. The latter 'pragmatic view' of dealing with disasters has the natural hazard itself as the principal object, and often treats the range of underlying reasons for the dangerous situation as irrelevant and immaterial. The factors involved in linking root causes and dynamic processes to vulnerability are seen as too diffuse or deep-rooted to address. Those who suggest they are crucial may be labelled as unrealistic or over-political.[3]

Our view is that there is little long-term value in confining attention mainly or exclusively to hazards in isolation from vulnerability and its causes. Problems will recur unless the underlying causes are tackled. This perspective does not reduce the importance of technical or planning measures to reduce physical risks, it simply insists on concern for the deeper level. Disaster research and policy must account for the connections in society that cause vulnerability as well as the hazards themselves.

GLOBAL DYNAMIC FACTORS

There is a serious lack of analysis of the linkages between vulnerability and major global processes as root causes. For example, it is not possible to identify the precise manner in which urbanization increases hazard impact. This situation reflects the preoccupation of most disaster work with the hazards themselves, and we propose that the IDNDR should collect and analyse data to determine the nature of such links.

Despite the shortage of firm evidence, there is a consensus that, for example, urbanization has contributed considerably to the severe losses of certain urban earthquakes of recent years, that population increase is one of the reasons for rapidly rising casualty statistics as a result of droughts and flooding, and that deforestation increases flooding and landslide risk. We cannot make any proper claims that vulnerability produced by a range of social processes has been increasing along with these factors. But we consider that the analysis and discussion of this book strongly supports such a view. This shows that it is imperative that the links between unsafe conditions and global pressures be analysed and understood, so that resources and attention can be put into reducing the pressures that generate vulnerability.

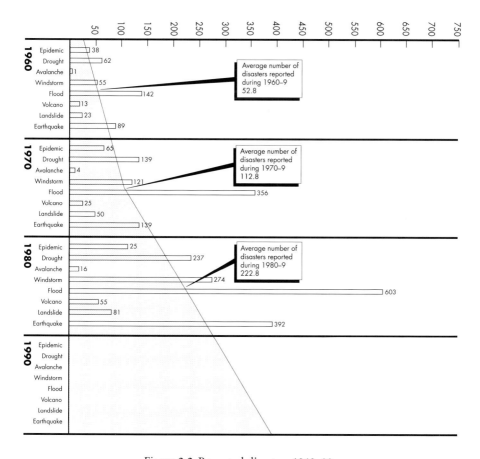

Figure 2.2 Reported disasters 1960–90

Source: Centre for Research in the Epidemiology of Disasters (CRED) 1991.

There is a general consensus in research on disasters that the number of natural hazard events (earthquakes, eruptions, floods, or cyclones) has not increased in recent decades. If this is valid, then we need to look at social factors that increase vulnerability (including, but not only, rising population) to explain the apparent increases in the number of disasters, in the value of losses and numbers of victims. Figure 2.2 shows the number of hazard events reported to have resulted in disasters for the last three decades. Some of the increase may be a result of better reporting and improved communications, or the incentive for governments to declare a disaster to try and win foreign aid. But the rising trend seems to be too rapid for these explanations alone, with a doubling time of around ten years.

A similar rising trend is suggested by the figures for economic losses and insured property losses compiled for the international insurance business (Figure 2.3). These are likely to reflect damage in industrialized countries much more than others, though part of the increase would be explained by the rise in capital-intensive insured property in newly industrializing parts of the Third World in the last two decades. These data are not a good measure of the number of disasters or people affected. This is not only because they refer to property, but also because one or a few large events can be very costly in financially-measured damage, while millions of people with little property can be severely affected by these or other events and remain unrecorded in such statistics.

Yet another approach is shown in Figure 2.4, again compiled for the insurance industry. This shows a rising trend in reported natural disasters similar to that of Figure 2.2, but with many fewer disasters in terms of the absolute numbers supposedly reported. There is no space here to go into the issue of the various sources of different sets of data. Our main point is to show that the trend is rising rapidly without there being evidence for any similar increase in natural hazards. But it should be noted that the data on disasters are riddled with problems as already mentioned, and that there is no common definition of what constitutes a 'natural disaster'.

It is also probably futile to try to quantify the increase in vulnerability worldwide. Yet there are many partial statistics (including reinsurance claims, the numbers of appeals for disaster assistance, and casualty estimates) that support a consensus among disaster workers that people's vulnerability is increasing (Shah 1983; Wijkman and Timberlake 1984; Drabek 1986; Berz 1990). Part II of this book suggests a wide variety of causal chains to account for specific vulnerabilities. However, among them there appear to be a number of root causes or global pressures which are particularly relevant in augmenting vulnerability.

At this stage, it is important to review in very broad terms how certain of these various dynamic pressures contribute to the increase in disasters. We have chosen six global processes for further attention: population growth, rapid urbanization, international financial pressures (especially foreign debt), land degradation, global environmental change, and war. It will also be noted that these processes are not independent of each other. They are themselves intricately connected in a series of mutually-influencing relationships that obscure causes and consequences.

Population growth

During 1987 the world population passed the 5 billion mark, yet only ninety years ago it was under 2 billion. Although the predictions of very rapid population growth forecast in the 1970s by the Club of Rome have not materialized (because of a slowing of growth in China and most industrialized countries), further rapid growth is regarded by demographers as inevitable.

32

It is now virtually certain from the number of people currently on the planet that there will be a further billion by the end of the International Decade for Natural Disaster Reduction (IDNDR) in the year 2000. The UN predicts a total of 6.1 billion for the year 2000, and a further 2 billion more people in the following twenty-five years, making a total of 8.2 billion by the year 2025. Thus in the space of thirty-five years there will be eight people in the world for every five there are at present, and 90 per cent of this growth

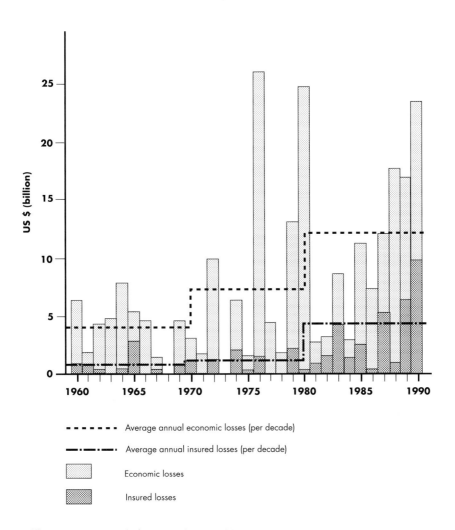

Figure 2.3 Economic losses and insured losses due to natural disasters 1960–89

Source: Sigma 2/90, Swiss Reinsurance Company 1990.

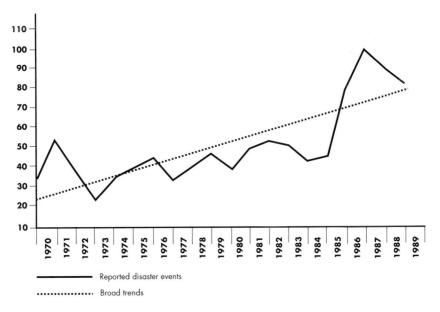

Figure 2.4 Number of natural disasters 1970–89

Source: Sigma 2/90, Swiss Reinsurance Company 1990.

will occur within developing countries, many of which are hazard-prone (United Nations 1986).

It is difficult to object to the idea that population growth is a significant global pressure contributing to increasing vulnerability. However, the link-ages remain uncharted, and share all the difficulties that exist in trying to explain demographic change more generally. For instance, since there is considerable debate about whether population growth is a cause or a conse-quence of poverty in the Third World (or a complex interaction of both), it is difficult to be certain about how to equate an increase in vulnerable people and disaster victims with population growth. Some of the implications of population expansion relative to disaster risks can more easily be related to different age groups (see Box 2.3), but we also need an analysis of the consequences of growth in numbers. This requires a better understanding of the linkages of population growth to disasters, and the nature of any causality involved. We would not deny that rising population is a com-ponent of increased vulnerability. But demographic processes themselves are largely a reflection of people's individual responses to the opportunities and uncertainties presented to them by broader economic processes. Therefore we would not want to accept a simplistic linking of population growth with vulnerability that suggests more people suffer more disasters because there are more of them in dangerous places. It is also necessary to explain why

34

people put themselves at risk. This itself is a process not explained by the increase in numbers alone, but by the differential access to incomes and resources in society. The apparently illogical behaviour of people who seem to have too many children in hazardous places can be seen to be more logical (if no less risky) in the context of the access model used in the next chapter. Is it significant that rapid population growth occurs in some countries with a long record of disasters?

Bangladesh, with its problem of land shortage, has a population of over 100 million that is growing annually at a rate estimated at between 2.7 and 3.5 per cent (doubling in under twenty years), with a land area of only 144,836 sq km (55,907 sq miles) (a similar land area to the state of Wisconsin). Land shortage is created by restricted access to land, a factor in many forms of vulnerability described in Part II. Eighty-five per cent of the population depend on agriculture, and between 40 and 60 per cent own no land (Boyce 1987). A vicious cycle is at work. People are likely to want large families where their survival strategy depends on wage-labour and diverse livelihood activities.

In Chapter 3 we show that livelihood strategies are the key to understanding the way people 'cope' with hazards. Unequal access to land and the resulting poverty and vulnerability of families is one of the factors that drives population growth (Hartmann and Standing 1989). Extreme forms of vulnerability are the result. Brammer has noted:

> Growing population pressure has increased the number of landless families . . . , increased the rate of rural–urban migration and forced increased numbers of people to seek living space and subsistence on disaster-prone land within and alongside major rivers and in the Meghna estuary.
>
> (1990a: 13)

The consequence of population growth in the context of highly unequal access to land (made worse by river erosion that takes land from one farmer and deposits it with another) is that more and more marginal land is being settled. This is particularly true of the low-lying islands (known locally as *char*) that emerge as a result of silt deposition in the river estuaries of the delta regions. This poses severe risks to the occupants from both cyclones and riverine flooding (see Chapters 6 and 7).

Whereas this situation is often considered hopeless from a 'technical' point of view, there are a range of *social* solutions that would both reduce the desire for large families and reduce disaster vulnerability. These solutions could include radical land reform, democratization of politics, empowerment of women, and the provision of adequate public services (health, communications, education). China, Sri Lanka, and Kerala State in India have all reduced population growth in this way (Hartmann 1987: ch. 14; Franke and Chasin 1989).

Box 2.3 Population growth, age structure, and vulnerabilities

The first concern is of the increasing proportion of young people in the world (M. Green 1979). Already in many developing countries as much as 50 per cent of the total population is under 15 years of age (compared with 20 per cent in industrialized countries). By the end of the century, the population under 30 years will increase by 500 million and constitute 60 per cent of the world population.

Although a high proportion of these children and teenagers engage in productive economic activity, it will be increasingly difficult to cater for their basic needs since a relatively small percentage of the adult population has to carry the responsibility for feeding, clothing, housing, and educating them. It is possible that this highly vulnerable segment of society could be neglected. Many may have no option other than becoming 'street children', forced to fend for themselves in hostile urban environments (Ennew and Milne 1989; Hardoy and Satterthwaite 1989).

This presents another type of 'disaster scenario' to the traditional hazards described in this book, one that would certainly increase the vulnerability of such children to a wide range of hazards. For example, street children may be more vulnerable to cholera because of the unsanitary conditions under which they live. If driven by poverty into the sale of sexual services, they are much more vulnerable to the risk of HIV infection (see Chapter 5).

Implications of this changing age profile are the need to focus disaster awareness programmes on to the needs of children, as well as the critical importance of making all school buildings resistant to hazards. But there is also a positive aspect to the young population profile and this is the potential of a very large, strong, energetic group of people under the age of 25. With the necessary political will, can we envisage the mobilization of predominantly young communities to protect their settlements against disaster risks? This implies the development of training, leadership, and suitable institutions for such a task.

At the other end of the age spectrum there is also a growing problem with the rapid increase in the number of old people. Within a ten-year period the world's elderly population will have increased by 43 per cent, with 70 per cent of this increase taking place in developing countries. Studies of disaster casualties have indicated that the young and the old are often most at risk. They are, for example, less mobile (capable of evacuation), more dependent, have less resistance to disease, and often command fewer resources. Increasing casualties in disasters can be anticipated in this age group. The implication is that specific risk-reduction policies will be needed to focus on the protection of the elderly, who have so far received minimal attention (ICIHI 1988: 16).

We argue that reducing disaster vulnerability and slowing global population growth share a means as well as an end. A 'safe environment' is the goal of both, but it is also the means. Reducing vulnerability to disasters will be shown to be tied up with increased resource access and empowerment of marginal groups. These are means to a more secure environment, but they are also the means to secure livelihoods that do not require large families.

During the IDNDR one of the greatest challenges will be to focus creative energy and resources, not on how to reduce population growth, but rather on tackling a more pragmatic question: how can much larger concentrations of people be adequately fed and safely accommodated within already congested disaster-prone rural and urban environments? As well as the occupation of dangerous rural areas the population increase will inevitably result in greater numbers of people moving to dangerous urban locations, which leads this discussion to the subject of another global pressure, that of rapid urbanization.

Rapid urbanization

This appears to be a key factor in the growth of vulnerability, particularly of low-income families living within squatter settlements (Davis 1987; Hardoy and Satterthwaite 1989: 146–221). The urbanization process results in land pressure as migrants from outside move into already overcrowded cities, where the new arrivals have few alternatives other than to occupy unsafe land (Havlick 1986). But the risks from natural hazards are only a part of the dangers these communities face; there are often far greater and more pressing 'normal' risks of malnutrition and poor health (Richards and Thomson 1984; Pryer and Crook 1988; Cairncross, Hardoy, and Satterthwaite 1990b).

Maskrey studied the vulnerability of squatter settlements to disaster risks with a team from a voluntary agency (Maskrey and Romero 1983). They analysed the shanty towns of Lima (Peru), which contain 40 per cent of the city's population (4.6 million in 1981, up from only 645,172 in 1940). This study revealed that in Lima, in an 'earthquake of magnitude 8.2 Richter over 26,000 dwellings would be destroyed or made unusable and about 128,000 would be left homeless' (Maskrey 1989: 7–8).

The high proportion of slum dwellers in Lima is not unusual in Third World cities. Similar ratios can be seen in other hazard-prone cities. For example, in Manila (the Philippines) the inhabitants of squatter settlements constitute 35 per cent of the population (over 1.5 million people in 1972), many living in areas subject to coastal flooding. In Bogota (Colombia) 60 per cent of the population (over 1.3 million people in 1969) live on steep slopes subject to landslides induced by heavy rains or earthquakes. Calcutta is subject to flooding and cyclones; here two-thirds of the population (over 5.3 million in 1971) live in squatter settlements (Donohue 1982).[4]

Hewitt examined the literature on earthquake impacts, and found that urbanization was closely related to damage to new multi-storey construction and in the concentrated poor housing of squatter settlements: 'where older sections of cities are run-down, often they have become slums that modernization passes by. Here, even once solid buildings are weakened by neglect and decay to become death traps in relatively moderate earthquakes' (Hewitt 1982: 21–2). This situation was typified in the earthquake that severely affected decaying inner-city tenements in Mexico City in 1985 (Cuny 1987), and is discussed in more detail in Chapter 8.

Maskrey has argued that the inhabitants of such critical areas

> would not choose to live there if they had any alternative, nor do they deliberately neglect the maintenance of their overcrowded and deteriorated tenements. For them it is the best-of-the-worst of a number of disaster-prone scenarios such as having nowhere to live, having no way of earning a living and having nothing to eat.
>
> (1989: 12)

Slum residents often incur greater risks from natural hazards (especially landslide or mudslide) as a result of having to live in very closely-built structures which can disturb natural land drainage patterns and watercourses (see Chapter 8).

The urbanization process not only magnifies the dangers of hazard events; it is in itself partly a consequence of a desperate migrant response to rural disasters. There is evidence from Delhi, Khartoum, and Dhaka (Bangladesh) that rural families who have become destitute as a result of droughts or floods have moved to these cities in search of food and work. Shakur studied the urbanization process in Dhaka. His household surveys revealed that

> the overwhelming majority of Dhaka squatters are rural destitutes who migrated to the city mainly in response to poor economic conditions (37 per cent) (particularly landlessness) or were driven by the natural disasters (25.7 per cent) (floods, cyclones and famines).
>
> (1987: 1)

Current projections indicate that within the coming ten years there will be twenty-two cities with populations in the 10–25 million range. Of these, fourteen are in the Third World, eleven being in hazardous zones. In all, thirteen of the twenty-two are prone to major hazards (Table 2.1). One of the most disturbing facts is that of these cities no fewer than seven are within high seismic risk zones. These cities contain large numbers of buildings of variable quality, many of them poorly constructed or badly maintained. The vast majority of deaths and injuries from earthquakes result from building collapse and damage.

There must be very high priority given to effective land-use planning controls (e.g. for steep slopes, low-lying areas, exposed coasts, and flood plains), well-enforced aseismic building codes and construction standards in

Table 2.1 Major cities at risk

City/conurbation	Population 1980 (millions)	Projected population 2000 (millions)	Hazard(s) to which exposed
Mexico City	14.5	25.8	earthquake
Tokyo–Yokohama	17.7	20.0	earthquake
Calcutta	9.5	16.5	cyclone; flood
Tehran	5.4	11.3	earthquake
Jakarta	6.6	13.3	earthquake; volcano
Rio de Janeiro	9.2	13.2	landslide
Shanghai	11.7	13.2	flood; typhoon
Delhi	5.8	13.2	flood
Dhaka	3.4	11.2	flood; cyclone
Cairo–Giza	6.9	11.1	flood; earthquake
Manila	5.9	11.1	flood; cyclone
Los Angeles	9.5	11.0	earthquake; landslide
Beijing	9.0	10.4	earthquake

these mega-cities. This is particularly so given that they are expanding rapidly, with the obvious risk of sloppy construction standards (Tyler 1990). In addition, two of these largest of cities (Cairo and Dhaka) are located on low-lying deltas, and will be seriously affected by any sea-level rise.[5] It is essential that governments recognize that vast cities pose enormous risks which require resources and a broad protection strategy.

Global economic pressures

A further global pressure on vulnerability to disasters involves the workings of the world economy. Since the Second World War the global economic order has changed rapidly. In particular, the pattern of financial relationships between the industrialized North and the Third World has altered with decolonization. Prices are falling for agricultural and mineral exports on which the Third World has traditionally had to depend. Meanwhile, prices of their imported energy and technology have increased. This created circumstances in which many Third World nations faced great difficulty in maintaining their balance of payments. In addition, the oil price rises of 1973 and 1979 led many countries to incur foreign debts. These were transformed into repayment crises especially by the rapid increases in interest rates in the late 1970s and early 1980s. In many African countries debt servicing alone (i.e., payments of interest and charges) amounts to 40–50 per cent of export earnings (George 1988; Onimode 1989; ROAPE 1990). Financial assistance flows into Africa (net of debt payment and repatriated profits) have declined steadily (Cheru 1989; Adedeji 1991), and in some cases are exceeded by debt and interest repayments. Foreign debt amounted to a very high percentage of annual GDP in many Latin American countries in 1985: 107 per cent in Bolivia, 99 per cent in Chile, 80 per cent in Uruguay, 77 per cent in Venezuela, and 73 per cent in Peru. The Latin American average was 60 per cent of GDP (Branford and Kucinski 1988: 9).

The outcome of this pressure has been to intensify the need to export at any cost. At the national level, this world economic situation has added pressure to exploit natural resources to the fullest extent possible to maximize exports. As discussed in the next section, such 'growth mentality' has resulted in degraded forests and soil that increase vulnerability to disasters (Tierney 1992).

During and since the 1980s many indebted countries agreed to IMF 'stabilization' and World Bank 'restructuring' policies, or initiated their own so-called structural adjustment programmes (SAP) that involve cutting public spending. As a result, services such as education, health, and sanitation are often reduced, and state-owned enterprises privatized (both these measures leading to unemployment), while food subsidies are reduced. The effects on welfare have been well documented (Cornia, Jolly, and Stewart 1987; Onimode 1989), but so far there has been little discussion of the effect of such programmes on disaster vulnerability.[6]

There can be very direct links between vulnerability and the operation of the global economy, as in a case from Jamaica (Ford 1987). Because of its foreign debt, the government of Jamaica intervened in the financial sector to try and reduce inflation and stimulate production. Interest rates went up to over 20 per cent, and home mortgage rates ran at between 14 and 25 per cent. These financial changes took place in a situation where the government enforced rent control and levied an import duty on construction material. There was therefore a rapid decline in residential construction. Consequently there was an immediate

> increase in vulnerability of a significant proportion of the urban population to hurricanes and earthquakes. This results from the fact that property owners faced with such high mortgage interest rates and little hope of recouping this by increasing their rents (due to the rent restrictions) simply ignore maintenance.
>
> (Ford 1987)

The impact of 'structural adjustment' on vulnerability went far beyond the issue of building maintenance. Because of the high cost of finance, builders tried to keep the cost of construction as low as possible so some small profit could be made. Again, safety suffered.

Health and education budgets are not the only ones to suffer cuts under SAPs. Even more crucial is the fact that the government's own programmes to introduce preparedness or mitigation measures were also cut as a result of the economic constraints. It would be difficult to determine whether the severe damage to Jamaica from hurricanes Gilbert in 1988 and Hugo in 1989 were made worse by the economic policies described above, but such potential connections are clearly possible. An additional irony in the Jamaican situation is that part of the foreign debt burden that caused the government to launch its SAP was due to loans used to pay for previous hurricane damage (see Chapter 7).

An estimated 50,000 children under 4 years old suffer from malnutrition in Jamaica (Oxfam 1988). More than one-third of the labour force earns less than £5 per week, while four times this sum is needed to feed an average family. In Chapter 5 we argue that such a weak nutritional (and therefore health) status of a population contributes to other forms of vulnerability in the long run. If the Jamaican debt burden has had a negative impact on the poor, it is affecting an already impoverished people with a considerable proportion vulnerable to local hazards.

The World Bank, and especially the small number of rich Northern member countries with a majority of votes, together with various multilateral regional development banks, are neither blameless nor ignorant of this situation. There has been a great deal of criticism of Bank policy on the debt issue.[7] It has led to some changes, including the establishment of a 'Green Fund' for environmental restoration, and loans for additional 'safety nets' such as nutritional monitoring and supplementation during the period of maximum SAP-induced hardship. There has been some concern about debt reduction (in addition to rescheduling) through such mechanisms as debt-for-nature swaps. All these could have a favourable effect on vulnerability, yet the net effect of SAPs is still likely to be an increase in the number of people at risk.

Land degradation and environmental losses

Another significant global dynamic pressure is destruction of forest, soil, wetlands, and water sources. This is often closely linked with the debt question, since land degradation may result from national policies favouring export production. In order to service debt, new lands have been cleared (e.g. in Brazil, the Philippines, Indonesia, and many African countries) for ranching or commercial cropping. Coastal areas have been drained, mangrove forests cut, in order to accommodate the expansion of tourist hotels and other foreign installations that hold out the hope of hard currency earnings. Likewise, much forest has been destroyed by the timber industry in Asia and Africa, where uncontrolled cutting of high-value exportable hardwoods is another way debtor governments can pay.[8]

The connection between land degradation and unsafe conditions can be quite significant (Pryor 1982; Cuny 1983). Deforestation and soil erosion can increase hazard intensity or frequency in the long run. The connection between deforestation and slope stability, erosion and the risk of drought, and other issues will be discussed at various points in Part II of this book.

41

Also, Chapter 5 will call our attention to the fact that extinction of wild genes (sometimes called 'genetic erosion') can significantly increase future vulnerability to plant pests and diseases. Deforestation and wetland destruction are major factors in such genetic erosion, leading to the loss of many species, known and unknown.

Another important aspect of loss of species and genetic variation is the changes in cropping systems and especially the increasing tendency for farmers to use fewer varieties of crops. Modernization is accompanied by dietary change, with imported and processed food items replacing traditional varieties of grain, legumes, fruits, and vegetables. Farmers grow a more limited number of commercial crop varieties, and the traditional ones die out (Juma 1989). When biological hazards strike, there may be no resistant varieties (genetic ancestors of the affected crops) on which to fall back. The Irish 'Potato Famine' of 1846–8 is a classic example. Irish peasants simply did not have access to (or knowledge of) the South American tubers that might have been imported to improve the disease resistance of the land race. Today, the destruction of world environments is wiping out the wild ancestors of many crops altogether. In the 1970s farmers in the US were able to get hold of other seed sources when maize (corn) blight halved the yield of monocropped hybrids they had become reliant on. The next time the older varieties of maize may no longer be available as insurance because they could have been wiped out (Fowler and Mooney 1990).

There are other important connections linking environmental destruction with the global pressures already discussed. Population growth and urbanization increase the demand for energy and in many countries dams (often large-scale) are being built to produce electricity. These dams flood vast areas of forest and other lands, forcibly displacing the inhabitants. With the growing demand for wood and charcoal in urban areas these fuels are brought into South Asian and African cities from hundreds of kilometres away (Little and Horowitz 1987; Leach and Mearns 1989).

The physical growth of cities has caused the destruction of much coastal wetland. Such swamps are drained for living space, for urban-fringe gardening, for fish ponds or salt works. Mangroves are cut for building material. Chapter 7 will emphasize the importance of these wetlands as buffers against coastal storms (Maltby 1986).

Global environmental change

There is growing evidence of changes in the interacting systems of the atmosphere, hydrosphere, and biosphere as a result of the build-up of 'greenhouse gases' from atmospheric pollution (Liverman 1989). The dangers are that the changes will increase the intensity and frequency of climatic hazards, and enlarge the areas that are affected by them. It is not possible definitively to blame the powerful hurricanes Gilbert, Joan, and

Hugo (1988 and 1989), the record storms in Europe in the winter of 1989–90, and the Australian floods of 1990 on the 'greenhouse effect'. But global climatic change provoked by warming is predicted to increase the number and intensity of storms, cyclones, and to amplify the variations in precipitation over much of the earth's surface. The impact on livelihoods will be immense (especially for farming and fishing peoples), in addition to the dangers from any intensification of the hazards.

We have already mentioned the danger of rising sea-levels as a likely result of global warming. Other predictions speak of the destruction of livelihoods (and possibly also lives) of a possible 6 million farm-workers living in the fertile delta regions of India. The UN-convened Intergovernmental Panel on Climate Change reported in 1990, and predicted a rise in mean temperatures of between 1.3° and 2.5°C and a rise in sea-level of 10 to 32 cm (3.9 to 12.6 inches) by the year 2030. A rise of this level is likely to have a significant impact on low-lying areas of many islands, as well as the flood-prone delta regions such as Bangladesh and Guyana. In the Pacific, Tuvalu and Tonga may become uninhabitable (J. Lewis 1989; Wells and Edwards 1989) and coral atolls which are home to many people in the Pacific and Indian oceans would experience submergence or destruction from storms.

Our understanding of global warming, and its implications for increased vulnerability to coastal flooding, tsunamis, cyclone, and storm surges is at a very early stage. Too many of the gloomy predictions of islands being drowned suggest an inert population, passively awaiting such cataclysms. The evidence from similar contexts is to the contrary, with actively mobilized governments and concerned individuals taking action to protect their threatened landholdings.

It is hoped that the real mitigation against global warming will not require the building of dykes and sea walls (to treat a symptom), but rather will involve the concerted actions of the governments of the world to reduce the production of greenhouse gases (carbon dioxide, methane, etc.) without further delay in order to treat the cause (Bach 1990). The tendency towards heroic engineering responses is always with us, as we shall see in the case of flood prevention in Bangladesh later on (Chapter 6).

War as a global pressure

There will unfortunately have to be frequent mention of war in the case-study chapters in Part II. There have been more than 120 wars since the end of the Second World War (van der Wusten 1985). They have had disastrous consequences in their own right for the people caught up in them, but they have also influenced vulnerability to extreme climatic and geological processes. At the regional and local scale war has disrupted and degraded the environment in, for instance, Vietnam (SIPRI 1976) and the Gulf (Kemp 1991; Seager 1992). Bomb craters, burning of forest or wetlands, or poisoning with herbicide (SIPRI 1980; Westing 1984a, 1984b, and 1985) can either trigger extreme events (such as landslides) or remove people's protection from extremes (such as coastal mangroves as a screen against high winds). Unexploded land-mines deny people access to arable land, thus reducing food security. Many rural people in Angola, Mozambique, and Cambodia have lost limbs trying to farm in heavily mined areas.

The economic impact of war, especially so-called 'low intensity' or 'counter-insurgency' warfare, is very hard on isolated rural households, who are often highly vulnerable to begin with. Contending forces ebb and flow over such peasant lands, extracting rations or tribute, making life insecure.

The influx of refugees produced by war in a neighbouring territory can have an immediate and dramatic influence on vulnerability by suddenly raising the population density (Hansen and Oliver-Smith 1982; Jacobson 1988). Demands on local services and infrastructure increase, fuelwood and water needs must be met, often with damaging consequences for the local environment. This local population pressure can increase disaster vulnerability in its own right (see section on population growth above).

NOTES

1 This way of organizing proximate and ultimate causes has been used elsewhere (e.g. in explaining land degradation by Blaikie and Brookfield (1987); Blaikie (1985a, 1985b, 1989)).
2 In 1988 Pakistan had the second highest balance of payments deficit in the world of US$3.5 billion.
3 See, for example, Bryant (1991: 7–8), who labels those who might want to consider social processes as 'Marxist', almost as a form of denigration.
4 The figures here are rather dated, reflecting in part the lack of frequent censuses and the difficulties in collecting accurate information from unofficial residents.
5 They have received special attention in the UN Climate Change Study 1990. There are also major studies in progress on other cities highly prone to rising sea-levels, including Hamburg, Venice, and Boston (O'Neil 1990; International Centre 1989).
6 Related problems of environmental degradation have been raised. In a paper for a meeting of CIDIE (Committee of International Development Institutions on the Environment), hosted by the World Bank, S. Hansen noted that for a number of reasons SAPs

> often lead to a deterioration of the situation for those with the least resources to adapt to the changed economic circumstances. To the extent that poverty in many regions of the world is the primary cause for environmental degradation, increased poverty caused by structural adjustment policies can lead to further environmental damage.
>
> (1988: 7)

He went on to describe ways the Bank has found to 'protect those who are virtually unable to adapt and compensate for the adjustment hardship'.

These modifications of the SAP design are described by Stewart (1987) and Haq and Kirdar (1987). It remains matter of controversy whether they are sufficient to protect vulnerable people and fragile environments.

7 There are some authors who advocate non-payment of the debt (F. Castro 1984) on the grounds that they have been paid many times over by unjust terms of trade. Others support some kind of 'delinking' from the world economy in favour of regional self-reliance (Mahjoub 1990; Amin 1990a, 1990b). Some see the World Bank as well-intentioned but severely constrained in how it can 'reach the poor' (Ayres 1983). Others believe the Bank consistently favours the elites and oligarchies in the Third World, making loans that, when assessed from social and environmental points of view, are insanely destructive (Hayter and Watson 1985; Linear 1985). Hancock (1989) argues that the Bank and other major development agencies are merely clubs for a class of experts that profits from poverty and 'the aid business'. George (1988) shares many of these doubts about the Bank, but believes it is reformable, as do Hellinger, Hellinger, and O'Regan (1988), who would reform bilateral development agencies such as the US Agency for International Development (AID) with the good example of NGOs.

8 It is clear than many of these damaging activities predate the debt crisis; the argument is that the response of governments and entrepreneurs to the priority for exports has intensified them. However, it is possible that the intensification of deforestation, for example, does not earn foreign exchange for the government to repay debt: in some circumstances individuals and enterprises control foreign earnings and syphon them off out of the country ('capital flight') without any benefit to the economy. There is also a serious corollary of this: that reduction of the debt burden may not alleviate the destruction of forests or other resources, since the motivation for damage is not always to service the economic problems of the nation.

3

ACCESS TO RESOURCES
AND COPING IN ADVERSITY

ACCESS TO RESOURCES

In the last chapter, we argued that disasters must be analysed as the result of
the impact of hazards on vulnerable people. We suggest two frameworks for
explaining this relationship between natural events and the social processes
that generate unsafe conditions. The first is the pressure and release (PAR)
model, which is designed to show in simple diagrammatic terms how
vulnerability can be traced back from unsafe conditions, through economic
and social pressures, to underlying root causes. This chain of explanation is
an analytical tool, subject to a number of inadequacies which we have tried
to illustrate.

One of its weaknesses is that the generation of vulnerability is not
adequately integrated with the way in which hazards themselves affect
people; it is a static model. It exaggerates the separation of the hazard from
social processes in order to emphasize the social causation of disasters. In
reality, nature forms a part of the social framework of society, as is most
evident in the use of natural resources for economic activity. Hazards are
also intertwined with human systems in affecting the pattern of assets and
livelihoods among people (for instance, affecting land distribution and
ownership after floods).

To avoid false separation of hazards from the social system, we propose
a second dynamic framework called the 'access' model. This focuses on
the way unsafe conditions arise in relation to the economic and political
processes that allocate assets, income, and other resources in a society.
But it also allows us to integrate nature in the explanation of hazard
impacts, because we can include nature itself, including its extremes, in
the workings of social processes. In short, we can show how social systems create the conditions in which hazards have a differential impact on
various societies and different groups within society. Nature itself constitutes a part of the resources that are allocated by social processes, and
under these conditions people become more or less vulnerable to hazard
impacts. In this chapter, the concept of 'access' to resources is explored

in a more formal way, and the model within which it can be understood is fully developed.

The concept can be illustrated with a narrative taken from the work of Winchester (1986, 1992), which analysed the impact of tropical cyclones (hurricanes) in coastal Andhra Pradesh (southeast India).[1] Cyclones in the Bay of Bengal periodically move across the coast and strike low-lying ground in Andhra Pradesh. They sometimes cause serious loss of life and property, and disrupt agriculture for months or even years afterwards. The damage is done by very high winds and often a storm-surge, followed by prolonged torrential rain. Let us compare how the cyclone affects a wealthy and a poor family living only 100 metres apart.

The wealthy household has six members, with a brick house, six draught cattle, and 1.2 ha (3 acres) of prime paddy land. The (male) head of household owns a small grain business for which he runs a truck. The poor family has a thatch and pole house, one draught ox and a calf, 0.2 ha (half an acre) of poor unirrigated land, and sharecropping rights for another 0.1 ha (quarter acre). The family consists of husband and wife, both of whom have to work as agricultural labourers for part of the year, and two children aged 5 and 2. The cyclone strikes, but the wealthy farmer has received warning on his radio and leaves the area with his valuables and family in the truck. The storm surge partly destroys his house, and the roof is taken off by the wind. Three cattle are drowned and his fields are flooded with their crops destroyed. The youngest child of the poor family is drowned, and they lose their house completely. Both animals also drown, and their fields are also flooded and the crops ruined.

The wealthy family return and use their savings from agriculture and trade to rebuild the house within a week (cost 6,000 rupees). They replace the cattle and are able to plough and replant their fields after the flood has receded. The poor family, although having lost less in monetary and resource terms, cannot find savings to replace their house (cost Rs 100). They have to borrow money for essential shelter from a private money-lender at exorbitant rates of interest. They cannot afford to replace the cattle but eventually manage to buy a calf. In the meantime they have to hire bullocks for ploughing their field, which they do too late, since many others are in the same position and draught animals are in short supply. As a result, the family suffers a hungry period eight months after the cyclone.

This anecdote illustrates how access to resources varies between households and the significance such differences in access have for potential loss and rate of recovery. Those with better access to information, cash, rights to the means of production, tools and equipment, and the social networks to mobilize resources from outside the household, are less vulnerable to hazards, and may be in a position to avoid disaster. Their losses are frequently greater in absolute terms but less in relative terms, and they are generally able to recover more quickly. As Rahmato (1988) has said: 'It is in

47

the years of recovery that the seeds of famine are actually sown.' In this illustration the seeds of further hardship, maybe starvation, have been sown for the household with poor access to resources, but not for the other.

This example helps to demonstrate the arguments of the first two chapters to show in general terms why variations in vulnerability to hazards are crucial in differentiating the level of impact on different groups of people. In general, rich people almost never starve, may avoid hazards completely or recover more quickly from events which are disastrous for others. A major explanatory factor of disasters is the distribution of wealth and power, because these act as determinants of the level of vulnerability of different people. We need to understand in detail how this distribution is structured and how it turns some natural events into disasters for some people. The idea of 'access' (especially to resources) is central to this task.

Access involves the ability of an individual, family, group, class, or community to use resources which are directly required to secure a liveli-hood. Access to those resources is always based on social and economic relations, usually including the social relations of production, gender, ethni-city, status, and age. This means that rights and obligations are not equally distributed among all people. Owners of land control that land and the crops it produces, even when it is worked by others. They can adapt the pattern of labour and other inputs as well as outputs to changing situations, particu-larly those which arise in the aftermath of a hazard. Gender is a pervasive division affecting all societies, and it channels access to social and economic resources away from women and towards men. Women are often denied the vote, the right to inherit land, and generally have less control over income-earning opportunities and cash within their own households. Normally their access to resources is inferior to that of men.

Since our argument is that less access to resources, in the absence of other compensations to provide safe conditions, leads to increased vulnerability, we contend that in general women are more vulnerable to hazards. Although the issue of gender will continue to appear throughout much of the book, we have chosen to investigate the relationship of gender and disaster in conjunc-tion with other variables affecting access (especially class, age, and ethnicity), rather than dealing with it in isolation. This is partly in recognition of the almost complete lack (in English-language sources we have used) of specific studies of gender and vulnerability in disasters.[2]

Resources required for people's livelihoods are rarely spread evenly in geographical space. Access may therefore also have spatial as well as political–economic dimensions. It may be too expensive (in all senses of cost) for someone to escape from a home or workplace which has inadequate livelihood resources, to travel to resources which are immovable and have a specific location. In most situations, the spatial inequality of access is a reflection of economic and social inequalities. There are many cases in which people who have little access to resources try to improve their situation by moving to new places (cities to seek work or to beg, or flood plains to farm) which are themselves more hazard-prone. Recovery from a hazard impact may similarly be impaired by spatial inaccessibility. For example, relief camps in famine areas may be too distant; unpolluted water sources unreachable in an area of flooding or earthquake. Looking at flooding in Bangladesh, Khan (1991: 340) has analysed 'previously unrecognized implications of the location biases for employment, shelter, access, and the utilisation of flood shelter by the powerless people'.

Many explanations of social change involve understanding how access to resources is determined in society. The way in which access changes over time, and the implications of this for different people, is also crucial. Long-term or intergenerational social changes must be analysed rather than just the immediate impact of a relatively sudden event like a famine or an earthquake. Access also characterizes the daily process of earning a living in normal conditions, under which each person has a different set of resources, and therefore has a different range of constraints and choices of livelihood, commensurate with those resources. It is out of 'normal' life that the social conditions for disasters emerge.

THE FORMAL FRAMEWORK: THE 'HOUSEHOLD' SUBMODEL

We will assume that the people we are concerned with in analysing vulnerability are members of economic decision-making units. These units can sometimes be called 'households', or 'hearth-holds' (Ekejuiba 1984; Guyer and Peters 1984), that is, those who share common eating arrangements which coincide with production units. Admittedly, there are cases where it is difficult to distinguish households at all.[3] However, such examples apart, it is usually possible to identify units which share labour and other inputs and consume together under one roof (or compound). We shall label these units 'households', each having a range or profile of resources and assets that represents their particular access level (box 2a in Figure 3.1). These may include land of various qualities, livestock, tools and equipment, capital and stock, reserves of food, jewellery, as well as labour power and specialist skills (box 2b in Figure 3.1). Non-material 'resources', qualities, or qualifications such as gender, membership of a particular tribe or caste should also be included. These are personal attributes and are not always unequivocally resources, since they may be essential prerequisites for some livelihood opportunities but bar the holder from others.

Access to these resources is secured through rights (e.g. property rights, rights accruing to women in marriage, and others sanctioned by law or custom). These rights may change, of course, particularly after the shock of disaster, so that the physical resource may still exist, but some individuals may no longer have access to it, or others may have greater access.

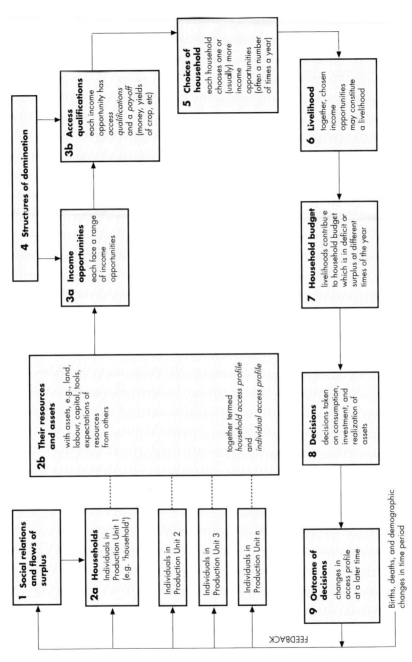

Figure 3.1 Access to resources to maintain livelihoods

Each household makes a choice (or is constrained) to take up one or more income or livelihood opportunities (box 3a). In rural areas, most of these will be the growing of different crops, or pasturing animals, while in urban areas there exists a wide range of opportunities including such things as petty-trading, working in a factory, casual labour, or domestic work.[4] Each income opportunity has access qualifications. This is defined as a set of resources and social attributes (skills, membership of a particular tribe or caste, gender, age) which is required in order to take up an income opportunity (box 3b).

Some income opportunities have high access qualifications such as capital, rare skills, or costly physical infrastructure, and therefore bar most from taking them up. As a result, they typically provide the highest returns. Others are much less demanding (e.g. casual labouring, which requires only an able-bodied person available at the point of employment), and these are oversubscribed and usually poorly paid.

Each income opportunity has a pay-off in terms of physical product, money, or other services. The mechanisms which set pay-offs for different incomes are of crucial importance in studying disasters, principally because they are apt to change radically, and reduce the pay-offs to some income opportunities, leaving individuals without alternatives.

Access to all the resources that each individual or household possesses can collectively be called its access profile. This is the level of access to resources and therefore to income opportunities, with some households having a much better choice than others (box 5). Those who possess access qualifications for a large number of income opportunities have a wide choice, and choose those with high pay-offs or low risks. They also have flexibility in securing a livelihood under generally adverse conditions, command considerable resources, have reserves of food, and can be said to have a well-resourced profile. On the other hand, those whose access profiles are limited usually have little choice in income opportunities, and have to seek the most oversubscribed and lowest paying ones, and have the least flexibility in adverse conditions. Those with a limited access profile often have to combine a number of income opportunities at different times of year. Some may be insufficient to provide a livelihood, and are only seasonally available or unreliable because others are competing for a limited number of places.

Each individual or household therefore makes choices, typically during key decision-making times in the agricultural calendar, or more irregularly under urban situations. The resulting bundle of income opportunities, together with the satisfaction of such needs as water and shelter, can be said to constitute a 'livelihood' (box 6). It is the sum of the pay-offs of its constituent income opportunities.

Some households structure their income opportunities in such a way as to avert the risk of threatening events such as drought, flood, or pest attack. They also employ survival strategies and coping mechanisms once that event

51

has occurred, though this usually involves an element of physical or institutional preparation. Grain must be stored, cattle numbers increased in good years to protect the reproductive capacity of the herd in bad years. A network of obligations and rights is built up in the form of institutions which deal with these events and aim to prevent them from becoming disasters. Therefore the choice of income opportunities is governed not only in terms of shorter-term objectives but sometimes with a longer view.

It is also important to map the spatial dimension of where, when, and how households spend their time earning their livelihoods. This is because natural events that trigger disaster have important spatial and temporal dimensions. Earthquakes, tsunamis, landslides, floods, and mudflows only affect particular locations, and disasters associated with extreme meteorological events such as high winds and heavy rainfall tend to occur during particular seasons. Cyclones coincide with the paddy harvest in Andhra Pradesh, with the result that migrant workers find themselves in hazardous places at hazardous times.

Another very different example is afforded by the AIDS pandemic in eastern and central Africa. Certain places (towns, trading centres, and settlements along roads) have high rates of HIV sero-positivity, exposing the population to very high risks of infection. Under these circumstances it may be very difficult for some people to avoid infection. For example, a wife would find it very difficult to refuse marital sex despite possible fatal consequences. Thus another element in the access model is the 'space–time mapping' of daily life (Carlstein 1982).

The flows of income then enter the household as a range of goods and cash: wages, grain, remittances from absent household members, profits from commerce or business, and so on. A household budget can be constructed in which expenditures and income are listed; the account accumulates, is in equilibrium or in deficit (box 7). On this basis decisions are made about how to cope with deficits, save or invest any surpluses, and what forms of consumption should occur (including arrangements for marrying children, having babies, migrating). If in surplus, the household may decide to invest and improve its access to resources in future. If the account is in deficit, consumption will have to be reduced, or assets disposed of, or they will have to postpone and possibly increase the deficit in the long term by arranging a consumption loan (which may be inadequate). The outcome of these decisions will result in a change in the access profile of each household in the next period. These will in aggregate alter the flows of surplus between groups and households (box 1) and may alter the social relations between groups, so that in the next round the households are in different relations to each other and larger-scale structures, and enter into box 2b with different access profiles. A disaster can cause a sudden deficit in the household budget, making that household more vulnerable to the next event (or disaster). This has been called the 'ratchet effect' (Chambers 1983) and is discussed in more detail later when we deal with issues of recovery and vulnerability reduction (Chapter 9).

THE FORMAL FRAMEWORK: HOUSEHOLDS IN SOCIETY

The outline of the 'household model' above may seem rather an economistic treatment of access to resources. We need to include more discussion of 'the rules of the game' or social transactions, and specifically of rights and social expectations which may give people access to resources. In abstract terms, the distribution of access to resources can be broadly accounted for by an analysis of class and power. The dominant relations of production and flows of surpluses are the main explanation of access to resources. Changes in the political economy at the level of 'root causes' in the PAR model are slow moving but can, as a result of revolution or major realignment in the balance of class forces, lead to fundamental shifts in the access to resources and in the character of disasters.

What is of more practical importance are the structures of domination and resource allocation (box 4 in Figure 3.1), or the 'rules of the game'. Between individuals within a household, these involve the allocation of food, who eats first, who will have to absorb consumption cut-backs in times of dearth, or who receives medical treatment. Gender politics within the household are of great importance here, and show how inadequate it is to treat the household as a homogeneous unit. As Rivers (1982) and Cutler (1984, 1985) among others have pointed out, women and children frequently bear the brunt of disasters because of the discriminatory allocative power of male members of the household while in refugee camps.[5]

Among family and kin, an important aspect of resource allocation is embodied in a range of expectations and obligations involving shelter, gifts, loans, and employment. Often these linkages reflect and reproduce the structures of domination of households and society in general. Between classes and groups, transactions include patron–client relations, taboos, untouchability, the sexual division of labour outside the home, sharecropper–landlord relations, and rules about property and theft, among many others.

Many of these transactions form an important basis of mutual help or individual survival in times of crisis, and therefore can be looked on as additional elements in an individual's or household's access profile (box 1). However, the rules governing these transactions change, often very fast, in the face of social upheaval such as war, famine, or pandemic. Usually this means a reduction in obligations and therefore in 'income opportunities' for the receivers of goods and services. In a few circumstances, new opportunities open up. For example, in times of extreme hunger, theft may be sanctioned, and grain stores may be broken into.

Markets are another set of social transactions that allocate resources on the basis of price. Their behaviour is crucial in the relative worth of people's resources, and in governing their household budgets. There is a good deal of research into the behaviour of markets, particularly preceding and during famines. Prices of essential goods and services often rise after sudden disasters where immediately-available food, shelter, clothing, and medical supplies are destroyed and transport to bring in replacement supplies is disrupted. The behaviour of traders in essential commodities is crucial as the next chapter on famine will show. These rapid changes of the rules of resource allocation bring us to the issue of analysing the dimension of time in disasters.[6]

Access to common property resources (CPRs) is also of great importance to household livelihood and vulnerability. At various times, in various broader social and economic settings, a wide range of physical resources may have been excluded from private or state ownership, and exist as property in common. These resources might include trees, pasture, ground- or surface-water, wildlife, marine resources, and arable land, depending on the region and its history. In some places some of these may be set aside for common management and use by a group larger than the household. Rules governing access to CPRs are highly localized and complex (Jodha 1991), and will be observed in many situations described in Part II.

Research on famine has led to the development of other concepts related to the idea of access. Most notable is Sen (1981), whose concept of 'entitlements' in relation to food and hunger has affinity with the notion of access. This involves the set of resources or livelihood opportunities which may be used to produce food or procure it through various forms of exchange. His formulation is similar to the concept of access in many ways, though it is more specific; his ideas are discussed more fully in Chapter 4. Swift (1989) has also put forward an approach in which assets, production, consumption, and exchange are interrelated and alter in ways that create problems for vulnerable rural populations. Watts (1983b) has taken a similar approach, and, while commenting on Sen (1981), stresses the importance of political power in gaining access to resources.

Our own model is derived from rural economic analysis (see Box 3.1) which identified the way access to resources changed over time for various different economic and social groups. The framework analyses the longer-term situation of populations subject to natural hazards, and examines the reasons why some people (differentiated by gender, age, status, class, etc.) are more severely affected in disasters than others. Secondly, it employs an iterative mechanism (incorporates cycles of change over time) and therefore can examine the preconditions, impacts, and after-effects of a disaster. If needed, the impact of a series of disasters on particular groups can be examined over a longer period. Some groups of people suffer from repeated disasters with the result that their access to resources is progressively reduced, making them more prone to disaster and less able to recover before the next hazard occurs. Thirdly, a focus on specific and quantitative measures of access to resources adds precision to a number of terms used in the treatment of disasters, which have been subject to a great deal of inexact usage. Such terms as vulnerability, disaster-proneness, disaster protection, disaster prevention have to be precisely defined and measured in a framework of this sort.

INCORPORATING CHANGE OVER TIME

Time is 'of the essence' in the understanding of disasters. So far time has only been treated in the sense that the framework permits the succession of events in a process to be analysed, allowing for people's decision-making actions (such as the timing of actions like planting crops, selling assets, migrating and so on). The importance of time in understanding disasters lies in the frequency of the event, when the disaster occurs (time of day, season), and in the stages of the impact of the disaster after the hazard has occurred.

It may be said that disasters do not happen, they unfold. This may be glaringly obvious in the case of 'slow-maturing' (slow-onset) disasters such as famine, the even slower AIDS pandemic, or ozone depletion, processes which unfold over a period of perhaps thirty to eighty years or more. But it may seem inapplicable to sudden hazards such as tsunami, bush or forest fire, earthquake, or some floods. However, even in these cases, the preconditions for disasters ('root causes' and 'dynamic pressures' in terms of our PAR model) have been forming over a long period. Indeed, Oliver-Smith (1994) treats the Peruvian earthquake of 1970 as having 'root causes' that reach back 500 years to the Spanish conquest of the Inca Empire and the ensuing decay of Inca methods of coping with environmental risk (see Chapter 8). Our treatment of the 1985 earthquake in Mexico City in Chapter 8 follows a similar line of thought.

It is therefore important to give a 'temporal frame' to our access model, so that the consequences of timing in hazard impacts can be understood. Although the timing of hazards may be random, the 'temporal frame' of people earning their livelihoods and living their daily lives is not. For the shortest time-frame, the time of day or night at the onset of sudden hazards can be important. Ninety per cent of all people killed in earthquakes while occupying buildings die at night (see Chapter 8). The day of the week (particularly market days, rest or holy days) is also relevant in terms of possible concentrations of people.

Seasonality is one of the most important rural time factors. Chambers, Longhurst, and Pacey (1981) and Chambers (1983) have highlighted the impact of seasonality on health, nutrition, and people's capacity for hard work in the 'normal' annual cycle. The coincidence of a sudden hazard with the 'hungry' season (usually the wet season) when labour demands are

---

Box 3.1   Using access models in the real world

Our development here of the access model derives from the work of a team which included Blaikie in Nepal in the 1970s (Blaikie, Cameron, and Seddon 1977, 1980), combined with the subsequent analyses of a number of others in fields related to hazards.[7] Part of the work in Nepal involved a simulation model based on empirical data collected from a survey of 667 rural households in Nepal. It was designed to predict the changing fortunes of existing households under different scenarios for five, ten, and twenty years ahead. (A scenario is a hypothetical set of assumptions about the change of different variables in the future which govern the outcome of the mechanisms under analysis.)

Each scenario was constructed from different assumptions regarding environmental degradation and crop yields, availability and prices of non-traditional agricultural inputs, different levels of population growth, and so on. Existing access to resources by each household was established by household production and consumption surveys, and analysis of existing structures of power and resource allocation. The physical resources included amounts of land of different quantities, labour, tools and productive equipment, livestock, forest products (fuel, fodder, building materials, etc.). The existing structures of power and resource allocation concerned the way in which the labour market operated, wages and conditions of employment, the market for commodities, discrimination against certain castes or tribal groups in access to jobs both in the agricultural sector and outside.

The model simulated the future of different households from an empirically verified present, and various 'rules of the game' were created to govern accumulation, sale of assets to cover losses or food shortfalls, and strategies of accumulation and investment, or loss and disinvestment. Households which became vulnerable to food shortages could be identified, and adaptive and coping mechanisms were simulated in the computer program.

---

highest, food reserves lowest, and some major diseases most prevalent, can produce a much more severe disaster impact. The build up of famines may have a seasonal element, in that crop failures (or a number of successive ones) are sometimes involved. Food prices as well as wage rates for agricultural work have important seasonal dimensions which other factors can make worse and precipitate as famine (see Chapter 4).

The stages of the impact of a disaster after a hazard event are fundamental. The various elements in the vulnerability framework (class relations, household access profiles, income opportunities, household budget, and structures of domination and resource allocation) each iterate at a different speed. Table 3.1 summarizes typical time periods of change and gives some examples.

There is a fundamental difference in time between sudden disasters and slow-developing disasters such as famine or pandemics (in which the most acute distress may extend over a period of months and years). In terms of

Box 3.1 continued

There is an interesting sequel to this simulation exercise. An opportunity arose to revisit the 100 most vulnerable households eight years after the initial fieldwork and simulation. While the model predicted quite well those households which had maintained their access to resources and productive capability, it did less well with regard to the very poorest households. The model was able to predict 'survival' while eight years later the reality was fortunately less dismal. The reason for this was that a whole variety of additional coping mechanisms and low-level social security networks had been in operation. We return to this phenomenon later in this chapter in the sections on coping.

This simulation has a number of important characteristics that should be considered further:

1 It identified an 'initial' state of unequal access to resources on the part of households as they pursued the earning of their livelihoods.
2 It predicted changes in access to resources through time and in relation to particular outcomes (e.g. environmental degradation, destitution, accumulation).
3 It involved choosing variables and 'rules of the game' (how social transactions were carried out) which were pertinent to the particular problem under study.
4 It was a dynamic model and iterated (proceeded through cycles of change) through time.

This analytical framework is not new, and has been used for many other predictive tasks. It is particularly useful for analysing disasters, since it focuses on the very issues we think important to understanding, preventing, and mitigating them.

their mortality and damage to homes and livelihoods, some sudden disasters can be perceived in terms of hours (e.g. 9 hours on 30 April and 1 May 1991 in Bangladesh) or a few days. Affected populations may rearrange their accepted pattern of responsibilities and rights and combine into completely unfamiliar groupings. On the other hand, slow-onset disasters require careful analysis of market behaviour over time, while the breakdown in households' access to resources and livelihoods may turn critical over a period of a few weeks as hunger and reduced capability to look for food or work set in.

THE ACCESS MODEL AND VULNERABILITY

As our previous discussion showed, vulnerability is a measure of a person or group's exposure to the effects of a natural hazard, including the degree to which they can recover from the impact of that event. Thus, it is only possible to develop a quantitative measure of vulnerability in terms of a

Table 3.1 Time periods for components of the access model

Component of the access framework	Typical time period of change after disaster	Examples
Class relations Change in political regime	Months or years	Nicaragua (1972) earthquake Portugal (1755) earthquake Ethiopia (1974) famine
Household access profile	Sudden, immediate impact Weeks Weeks or months	Loss of life and house Sale of livestock, jewellery Other assets sold
Income opportunities	Sudden if urban employment is disrupted Usually over months	Rural employment collapses due to drought, flood Taboo foods accepted
Household budget	Immediate impact in sudden-onset hazards Months	Cuts in consumption; reallocations by age, gender Food prices risen and famine
Structures of domination	Immediate impact in sudden disasters Months or years, with episodic food shortages and merchants' high prices	Sharecroppers refuse to give up landlords' share Famine

probability that a hazard of particular intensity, frequency, and duration will occur. These variable characteristics of the hazard will affect the degree of loss within a household or group in relation to their level of vulnerability to various specific hazards of differing intensities.

Thus vulnerability is a hypothetical and predictive term, which can only be 'proved' by observing the impact of the event when, and if, it occurs. By constructing the household access model for the affected people we can understand the causes and symptoms of vulnerability. This requires analysing the political–economic structures which produce the households' access profiles, income opportunities, and pay-offs (these structures are labelled 'class structure' and 'structures of domination and allocation' in the framework).

This implies that the question 'vulnerable to what?' is answerable only in the context of an actual hazard. But it brings up an important point: different people will be vulnerable in differing degrees to different hazards. There will be households which if vulnerable to one type of hazard are likely to be vulnerable to others. Typically, such people will have a poor access profile with little choice and flexibility in times of post-disaster stress.

Nonetheless, in the following chapters, it is necessary to specify to what disasters people are vulnerable. In the case of earthquakes, clearly the indicators of vulnerability will concern housing materials, income level, available spare time and skills to keep habitations in good repair, type of tenure (owner-occupier or rented accommodation if urban areas are being studied), location relative to zones of seismic activity, ground stability, and degree of support networks which could be mobilized after the event (see Chapter 8). In the case of drought, the set of indicators of vulnerability will be quite different, and will concern food, entitlements profile, availability and market behaviour, and the prospects for earning enough to buy food or exchange other goods for it. The time–space patterns of households will be important as in the earthquake case, but related to the spatial structure of markets and to crop or pastoral production.

THE ACCESS MODEL AS A RESEARCH FRAMEWORK

The formal access framework has so far been presented as an explanatory and organizational device. It is not a theory, although theories of disaster can be inserted into the general framework. For example, in the next chapter competing or partial theories of famine are seen to deal with different parts of the framework.[8] It draws attention to the socio-economic relations which cause disasters or allow them to happen. While it focuses on those at risk of disasters, it also includes the relations they have with others which keep them in that unfortunate state. It also allows for people's response to the situation, either by coping or by more active and permanent efforts to change those relations (see the sections on coping later in this chapter).

The access framework therefore does not include national policies or world systems in the way that the PAR model does. The impact of national and international processes can be incorporated in the model. Land reform, food policy, famine relief, food-for-work programmes, rural reconstruction programmes, laws governing urban property, and so on can appear in all the elements in the framework, but their causal impact would be treated as exogenous to it.

As a research design therefore, the framework is useful in charting impacts of policy, for identifying vulnerable populations, and predicting the probable outcomes of extreme natural events. However, the data requirements for using the framework as a research design would be very large. After all, it provides a general outline of the material conditions of life of a population, and most aspects of society are potential inclusions. Yet we believe that in use the number of factors to be incorporated would be restricted, because in use the framework would be informed by theory and a priori assumptions.

This would lead to the ability to choose the most significant factors, and permit selectivity in the use of the framework.

Some of the main criteria for making these choices can be suggested:

1 The researcher's emphasis on certain theories and priorities will determine what has to be modelled in detail. For example, if gender relations are empirically an important element in disaster impact and policy, then the individual rather than the household would be the unit of study and the main focus. If a researcher believes that a supply-side theory of famine requires attention, then those income opportunities in crop production would be emphasized with reference to drought or pest attack, along with other determinants of supply (e.g. the transport network).

2 The scale of the investigation will also be partly determined by choice of theory as is implied in (1) above. Individual, household, class or village, region, and nation are not so much alternative objects of analysis, but rather a series of conceptual limits that nest inside each other (like Russian dolls or Chinese boxes), the smaller scale enclosed by the next highest level and influenced by it. Nonetheless, the study will have to choose the major spatial frame appropriate to its purposes. If a seismic belt, a farming system, or an administrative area is chosen as the principal scale of the study, other scales can be sketched in through secondary data, rapid rural appraisal, and key informants.

3 The framework is principally an externalist approach, in that it imposes the researcher's own interpretation and perception of vulnerability, hazard, and risk. The possible victims and other actors have their own: as Chapter 4 shows, for example, 'famine' is perceived in a variety of ways which differ significantly from those of the world media, aid, or relief agencies (de Waal 1987). There is no reason why primary data collection of indigenous interpretations of events and processes cannot enrich and perhaps alter the framework.

4 Most studies do not examine vulnerability for its own sake, but assist in the prevention or mitigation of disasters. Therefore many variables mentioned here in the general framework will simply not apply to particular hazards or in particular situations.

The access model is used as a predictive and organizing device for this book. Only parts of it, at the discretion of the researcher or policy-maker, will be relevant in each case.

VULNERABILITY AND POVERTY

Vulnerability and poverty are not synonymous, although they are often closely related. Vulnerability is a combination of characteristics of a person or group, expressed in relation to hazard exposure which derives from the social and economic condition of the individual, family, or community concerned. High levels of vulnerability imply a grave outcome in hazard events, but are a complex combination of both the qualities of the hazards involved and the characteristics of the people. Poverty is a much less complex descriptive measure of people's lack or need. Vulnerability is a relative and specific term, always implying a vulnerability to a particular hazard. A person may be vulnerable to loss of property or life from floods but not to drought. Poverty may or may not be a relative term, but there are not varying 'poverties' for any one individual or family.

There are also policy reasons for keeping the two terms separate (Chambers, Pacey, and Thrupp 1989). Anti-poverty programmes are designed to raise incomes or consumption, while anti-vulnerability programmes aim to reduce the chances of a hazard having a serious effect, and to increase 'security'. For example, an anti-poverty programme was initiated in Turkey to support those affected by earthquakes by increasing their income opportunities so that they could make good their financial losses. Male members of households were given the opportunity to work as *Gastarbeiter* ('guest workers') in the then West Germany. They were able to save relatively large sums of money and thereby became less poor. They invested their savings in large and unsafe houses thereby defeating the long-term purpose of the programme, and increasing the vulnerability but reducing their poverty (see Chapter 8).

Chambers, Pacey, and Thrupp (1989) also make the same point when they say that there are trade-offs between poverty and vulnerability where poverty can be reduced by borrowing and investment but that the impact of these interventions may increase vulnerability. These contradictions are obscured if poverty and vulnerability are taken as being the same thing. This is not to deny that there is often a strong correlation between poverty and vulnerability, as the case-studies in Part II show.

COPING IN ADVERSITY

The access model provides a dynamic framework of socio-economic change, in which people of different identities (gender, age, seniority, class, caste, ethnic group) avail themselves of the means of securing their livelihoods and maintaining their expectations in life. The model implicitly, rather than explicitly, allows for people to develop strategies to try and achieve these ends. In this sense, the economic and social means to secure their livelihoods are not 'handed down' to them in an economistic and deterministic manner. People must not be assumed to be passive recipients of a profile of opportunities, hedged about by constraints of the political economy of which they are part.

On the contrary, the pattern of access in any society is subject to and the result of struggles over resources. The pattern of access is the outcome of those struggles between people of different gender, age, class, and so on.

They are a part of daily life and are pursued with ingenuity and resourceful-ness. In adverse or disastrous times people are stimulated by circumstances of desperation and loss. As Rahmato (1988) puts it, the measures which rural Ethiopian people have taken to enable them to live through the privations of the past two decades indicate ingenuity, strength of character, and an effective use of natural resources and communalism.

It has been said that official perceptions of 'disaster victims' usually underestimate their resources and resourcefulness. Perhaps one of the reasons for this is that indicators of vulnerability based on the measurement of resources are the more easily recognized by outside institutions. They are also more enduring and part of the observable socio-economic structure, while people's struggles and strategies to cope with adverse circumstances, particularly acute ones, are more ephemeral and change quickly (Corbett 1988). Therefore they remain unnoticed and understudied. It is the purpose of this section to focus on these strategies. Without a proper understanding of them, policy-makers are more likely to make stereotyped responses in both preventive measures of vulnerability reduction and relief work. Further, misdirected relief efforts may undermine rather than assist affected people in their attempts to help themselves in recovery.

Coping defined

'Coping' is the manner in which people act within existing resources and range of expectations of a situation to achieve various ends. In general, this involves no more than 'managing resources', but usually means how it is done in unusual, abnormal, and adverse situations. Thus coping can include defence mechanisms, active ways of solving problems, and methods for handling stress (Murphy and Moriarty 1976). 'Resources' in this book have been defined as the physical and social means of gaining a livelihood. This includes labour power, or as Chambers (1989: 4) aptly puts it able-bodiedness, or the ability to use labour power effectively. The more poor people rely on physical work, the higher the potential costs in physical disability.

Resources also include land, tools, seed for crops, livestock, draught animals, cash, jewellery, other items of value which can be sold, storable food stocks as well as skills. In order for tangible resources to be mobilized, people must be entitled to command them, and this may be achieved in many ways. These include using the market, the exercise of rights, calling upon obligations (of other household members, kin, patrons, friends, of the general public by appeals to moral duty, as in alms-giving), through theft, or even violence.

In many cases specialized knowledge is required with certain resources, for instance, in finding wild foods, or using timber for rebuilding, knowing the moisture capacity of certain soils, the likelihood of finding wage-labour in a distant city or plantation, or finding water sources. This knowledge is similar to that which supports 'normal' rural or urban life, and is passed from generation to generation. However the 'ethnoscience' essential for some coping behaviour can disappear with disuse or be rendered useless by rapid change (O'Keefe and Wisner 1975).[9] We return to this point below, in the discussion of coping.

Often it is assumed that the objective of coping strategies is survival in the face of adverse events. While this is indeed common, it masks other important purposes. These may be examined using Maslow's hierarchy of human needs (Maslow 1970). Such a hierarchy involves identifying distinct levels of needs, with each level incorporating and depending on the satisfaction of needs below them in the hierarchy.

The need for self-realization, involving the giving and receiving of love, affection, and respect might be said to be the highest in the hierarchy. A lower one, on which the former is founded, may be an acceptable standard of living. Lower ones still may include adequate shelter and food for healthy survival, while other needs near the bottom of the hierarchy will include minimum security from violence and starvation. Reviewing twenty years of work since Maslow, Doyal and Gough (1991) conclude that a 'core' of basic human needs can be identified, and that failure to satisfy these means that other needs cannot be met (see also Wisner 1988b).

In adverse circumstances, a retreat to the defence of needs that are lower in the hierarchy implies the temporary denial of those higher up. For example, the experience of extreme poverty can cause a loss of self respect and self-regard (de Waal 1989b). However, it is important not to oversimplify and overgeneralize the expectations and priorities in life of vulnerable people or those affected by a disaster. Oliver-Smith (1986b) has described very complex motives and ideals among survivors of a horrible earthquake tragedy (see Chapter 8). J.C. Scott (1990: 7) reminds us that 'slights to human dignity' can fester and emerge in surprising demonstrations of 'resistance' against authority. This is certainly relevant for disaster relief and recovery (Chapter 9).

Jodha (1991) surveyed people's own criteria of well-being (in this case no fewer than thirty-eight) in Gujarat, India, which attest to a complex set of priorities. Raphael (1986) analysed the psychological trauma of disasters and the adjustments made to loss, grief, and the impacts of dislocation. Coping in the face of adverse circumstances therefore may be seen as a series of adaptive strategies to preserve needs as high up the hierarchy as possible in the face of threat.

However, it is possible that what may be broadly termed 'disasters' forces a retreat down the hierarchy. For example, it may become necessary to engage in demeaning activities (and therefore to lose respect) in order to secure a minimum food supply. Certain activities may be proscribed or discouraged by membership of a social group, caste, or by gender. During

63

the drought of 1971–3, members of the Reddy caste in Medak District, India, were reduced to selling vegetables to earn a living, an occupation that was considered below their dignity (Rao 1974), while women not of the shoemaker caste were found making shoes during the 1966–7 Bihar drought (Singh 1975; quoted in Agarwal 1990). Despite the mutual economic and emotional support that they provide, families may have to break up for their individual members to survive. The survival of the individual in the short term may be the only attainable need and objective of coping.

Famine may be unique or at least extreme among disasters in often provoking social degeneration of this kind. For many years Quarantelli and his sociology colleagues have studied community responses to disasters such as earthquakes and floods. They find that emergent organization is much more common than social chaos, and that altruism and stoicism are more common than selfishness and panic (Quarantelli and Dynes 1972, 1977; Quarantelli 1978, 1984; Dynes, De Marchi, and Pelanda 1987).

Types of coping strategies

Crisis events occur from time to time in people's lives, as well as in the lives of whole communities and societies, in which case they are often called disasters. Such events call for the mobilization of resources at various levels to cope with their impact. When people know an event may occur in the future because it has happened in the past, they often set up ways of coping with it (Douglas 1985).

Such coping strategies depend on the assumption that the event itself will follow a familiar pattern, and that people's earlier actions will be a reasonable guide for similar events. Most disasters have such precedents, particularly in hazardous physical and social environments. However, some hazards have such a long return period that the precedents are imperfectly registered. There are also others which are unprecedented, like the AIDS pandemic, having no familiar pattern. If this is the case then coping strategies may not apply, and the decision framework (consisting of the social, economic, and natural environments) will not be relevant.

The assumptions on which people make their decisions therefore rest in the knowledge that, sooner or later, a particular risk will occur of which people have some experience of how to cope. On the other hand, people do not like conditions of uncertainty where there are no known and familiar ways (such as explicit systems of rights and obligations, providing safety nets and support groups) of coping with a particular event. Thus the unprecedented or unknown event creates a situation of uncertainty. The AIDS pandemic in certain areas of Africa, or calamities of exceptional severity (as in the case of what are known in Bengali as *mananthor*, or 'epoch ending' famines) are cases in point.

Almost all coping strategies for adverse events which *are* perceived to have precedents consist of actions before, during, and after the event. Each type of coping strategy is discussed and illustrated below.

Preventive strategies

These are attempts to avoid the disaster happening at all, and are called preventive action elsewhere in this book. Many require successful political mobilization at the level of the state. This is often easier in the immediate aftermath of a disaster, when public awareness is high and the political pay-off of government action is significant.

But preventive action at the individual and small group level is also important. It may involve avoiding dangerous time–spaces, such as offshore fishing in small open craft during the storm season, evading seasonal and/or altitudinal concentrations of disease vectors (e.g. malaria mosquito, tsetse fly), and choosing locations for housing that are less exposed to wind, flood, or mass movement of earth.

Impact-minimizing strategies

These are referred to elsewhere as 'mitigation', especially where they are the object of government policy. These strategies seek to minimize loss and facilitate recovery. The range of these strategies is enormous, and varies significantly between people with different patterns of access. However, two generalizations may be made. Firstly, the objective of many strategies is to secure needs quite low down the hierarchy, particularly if the risk is perceived to be damaging and probable. It may be preferable to improve access to a minimum level of food, shelter, and physical security than to increase income. This further underscores the important distinction between poverty and vulnerability made earlier.

Secondly, maintaining command of these needs in a socially and/or environmentally risky environment usually implies diversification of access to resources. In the terms of the access model, this involves broadening the access profile. This can be attempted in agricultural and pastoral production, in setting up non-agricultural income sources, and strengthening or multiplying social support networks.

Creation and maintenance of labour power

Having a large number of children is thought to improve security by increasing the possible future incomes to which the family may have access. Cain (1978) has argued that children are considered to be a less risky investment than land, and also that a reduction in the perceived risk of severe and acute deprivation brings about a change in the desired number of children. Cain compared two communities in Bangladesh and nearby in east

India. The higher level of risk of disaster in Bangladesh seemed to create a greater desire for children and more actual births. By comparison, in the Indian village the level of mortality, social security, and health provision was superior.

Building up stores of food and saleable assets

For those rural people who have access to land, a store of grain or other staple food is a most important buffer against expected seasonal shortages, as well as more prolonged periods of hardship. An accumulation of small stock and chickens is another shield (Watts 1983b). Pastoralists may follow a strategy of increasing herd size in years of good rains and grass availability (when calf birth rates rise and mortality falls) in order to maintain herd size in the inevitable bad years with high mortality (Dahl and Hjort 1976; Thébaud 1988; Odegi-Awuoundo 1990).

Diversification of the production strategy

Farming people are often regarded as risk-averse (in the sense of avoiding chances in cultivation that may bring higher rewards but with greater exposure to dangers).[10] Usually their production involves mixed cropping, intercropping, the cultivation of non-staple root crops, and use of kitchen gardens. This strategy often results in a 'normal surplus' in good years since it is planned on the basis of meeting subsistence needs even in bad (but not the worst conceivable) years (Allan 1965; Wisner 1978b; Porter 1979).

Planting a greater variety of crops has many advantages apart from providing the best chance of an optimum yield under all variations of weather, plant diseases, and pest attack. It represents one of the most important precautionary strategies for coping with food shortages (Klee 1980; Altieri 1987; Wilken 1988). Diversification strategies often make use of environmental variations, including farming at different altitudes, soils, or diverse ecosystems on the slopes of mountain systems.

Diversification of income sources

The entirely self-provisioning rural household is an ideal type, very rare in the world today. Even the most isolated people in the Amazon rainforest, the Andes, northern Quebec (Canada), or the Himalayas engage in production for sale. In addition, the remittance of income from wage-earners who have moved to distant cities, mining camps, or plantations is very important to rural livelihoods in many parts of the world. This is sometimes graphically demonstrated by the economic disruption and hardship caused when crises interrupt such systems, as with the hundreds of thousands of guest workers from Egypt, Bangladesh, and the Philippines who left Iraq in 1991 as a result of the conflict.

Non-farm income becomes even more important following disasters that temporarily disrupt farm and livestock production. Crafts, extractive enterprises such as charcoal-making, honey and gum arabic collection have often been noted in studies of drought-coping in Africa. Brewing beer is also an important income source, especially for women, and a reduction of the grain ingredients caused by drought can affect their income and nutrition (Kerner and Cook 1991; Murray 1981; Mbithi and Wisner 1973). A series of 'sidelines', sometimes illegal or quasi-legal (such as hawking on the streets without a licence), is a mainstay of 'normal' life for urban dwellers and even more important as a fall-back if employment is interrupted by an earthquake, flood, or mudslide.

Both production and income diversification strategies can be effective as coping mechanisms in the short run, while they undermine the basis of livelihood in the long run. Cannon (1991) discusses devegetation of the landscape in order to provide fodder for livestock in a drought. Charcoal-burning as an income source is another example. Both can lead to long-term erosion and desertification (Grainger 1990; O'Brien and Gruenbaum 1991).

Development of social support networks

These include a wide variety of rights and obligations between members of the same household (e.g. wives and husbands, children and parents), with the extended family, and other wider groups with a shared identity such as clan, tribe, and caste. Parents may try to make a strategic choice of marriage for their daughter or son into a comparatively wealthy family. This may increase their ability to call on resources in difficult times (Caldwell, Reddy, and Caldwell 1986).

Within the household and family, successfully securing resources in potentially disastrous times depends upon the implicit bargaining strength of its members, and of their 'fall-back' position (Agarwal 1990: 343), or 'breakdown' position as Sen terms it, if co-operation in this bargaining process should fail (Sen 1988, 1990). Women tend to lose these conflicts for scarce resources, and are affected by who eats first, the share of available food, and lack of access to cash earned by other family members (e.g. cash from casual male labour). The range of resources controlled by women, and employment opportunities open to them, tends to be more limited.[11] The disintegration of the family, and abandonment of women, children, and old people is the expression of the breakdown of such obligations.[12]

There are other forms of support based mainly on non-economic relations. Some writers term these the 'moral economy' (J.C. Scott 1976), and they include relations between patrons and clients, or between rich and poor in times of hardship. These offer a minimum subsistence and a margin of security, and constitute what Scott has called 'a subsistence ethic', based on the norm of reciprocity. Examples are legion, but it is widely reported that such obligations are being eroded.[13]

On the other hand, Caldwell, Reddy, and Caldwell (1986: 667) state that, at least for the aged in a period of extreme food scarcity in south India, 'the support system still worked well'. Gupta (1988) goes so far as to say that the continued existence of such support in present-day India is responsible for the retention of people in the countryside; people are discouraged from abandoning such local systems. In other cases these obligations of the wealthy are still upheld. For example, a case-study in Nepal found that the wealthy were prevailed on not to reduce daily wages for agricultural work, nor to sell grain outside the village at a profit, and to secure a loan from shopkeepers in the nearby bazaar for relending to the most needy villagers (Prindle 1979). In another tribal village (as opposed to the multi-caste one in the previous example) there was an expectation of gifts in times of hardship combined with a powerful ethic of equality, with surpluses being shared. Although reported to be in a state of subtle change, it is still largely operational.

There are also wider obligations of the whole community to assist and provide for those facing acute adversity. These include alms, for example, the giving of a grain tithe in some Muslim societies (Longhurst 1986: 30). *Meskel* is a form of community redistribution in parts of Ethiopia, where credit is given to the needy to celebrate the festival of this name, who thereby acquire food. Neighbourly assistance such as the rescue of trapped individuals inside collapsed buildings and the rendering of medical assistance are other examples. These are 'claims', as Swift (1989) calls them, alongside the other two broad categories of assets ('investments', both human and productive, and 'stores' of food, money, and 'stores of real value' such as jewellery).

It is probable that generally throughout the South such networks and moral obligations are in decline. In some areas more exploitative ones are superseding the old semi-feudal paternalistic pattern. This may involve provision of food on credit at usurious rates of interest, which exacerbate the 'ratchet effect' and increase vulnerability of deprived groups in the longer term (Chambers 1983). With the demise of such traditional systems, there is rarely any growth of state-run higher level social security alternatives.

Post-event coping strategies

Once the hazard which had been foreseen, understood, and prepared for, actually befalls a population, the precautionary mechanisms are put into practice. There are also others, which cannot be developed in advance, but which come into play after the hazard event. In these cases behaviour is not random, but draws on relevant knowledge and analogous precedents.

When there is a potential food shortage and possible famine, the period during which stress develops can be long, allowing for a succession of strategies. A number of studies have found similar sequences.[14] It is clear that a sequence of adaptations in consumption patterns is made very early when shortfalls in food are anticipated. These include substitutions of lower quality and wild foods (and 'famine foods') for possibly more expensive staples. Here the significance of common property resources for allowing access to these foods is important.[15] Wild foods also feature as famine foods in almost all parts of Africa (de Waal 1989a; McGlothlen, Goldsmith, and Fox 1986).

The next step involves calling on resources from others (usually family and kin) which can be obtained without threatening future security. This usually involves reciprocal social interactions, and avoids usurious rates of interest, and therefore preserves the longer-term access position of the individual or household. At the same stage, sources of household income other than the dominant one may be tapped, such as wage-labour, petty commodity production, or artisanal work. Sale of easily disposable items which do not undermine future productive capacity (e.g. small stock) may also take place. As the food crisis deepens, loans from moneylenders and the sale of important assets such as oxen for ploughing, agricultural implements, and livestock may have to be arranged. Finally, when all preceding strategies have failed to maintain minimum food levels, migration of the whole household to roadsides, towns, and possible sources of food often ensues.

COPING AND VULNERABILITY ANALYSIS

Coping strategies are often complex and involve a number of sequenced mechanisms for obtaining resources in times of adversity and disaster. They grow out of the recognition of the risk of an event occurring and of established patterns of response. They seek not just survival, but also the maintenance of other human needs such as the receiving of respect, dignity, and the maintenance of family, household, and community cohesion. Often outsiders are surprised by strategies that do not seem to try and maintain adequate food intake for a household (or perhaps different amounts for various members), but which instead are aimed at preserving the means for continuing the livelihood after the difficult period has passed.[16] Many of these strategies have been highly resilient to social and economic changes and are reported to be still functioning throughout the world.

Many strategies seem to be coming under a variety of pressures which have reduced their range and efficacy. Rights and obligations of assistance in adverse circumstances that are embedded in feudal economic relations have been eroded, and familial ties have weakened in many parts of the world, particularly where non-agricultural incomes and partial or temporary outmigration have become common. The penetration of the market has had a less definite impact on coping strategies, both improving some and eroding others. Also, pressure on natural resources and increased competition in times of scarcity have strained co-operative and reciprocal behaviour. Nonetheless, coping strategies of all kinds are crucial elements in understanding vulnerability and designing interventions which provide sustainable self-help solutions to recovery and future disaster prevention.

Throughout this book we try to signal the ways in which the 'people's science' or indigenous knowledge, that provides the basis for much coping behaviour, and patterns of coping themselves, interact with 'official' attempts at disaster prevention and mitigation. Sometimes a sensitive administration or a non-governmental organization has been able to build on such foundations. Many examples are provided by Maskrey (1989) and Anderson and Woodrow (1989) and others, and we will return to this glimmer of hope in Chapters 9 and 10.[17] More often than not, however, 'official' relief and recovery practice pays little heed to what the ordinary people do. The result is wasted resources, squandered opportunities, and a further erosion of vernacular coping skills.

NOTES

1 This type of hazard, and the Andhra Pradesh case, are dealt with at greater length in Chapter 7. Winchester's work (1986, 1992) is a valuable and rare example of a study of the actual operation of ideas of vulnerability in analysis of an actual sudden-onset disaster.

2 Women's lives are almost always more constrained and difficult than men's. Hazards feature among the difficulties, but are never alone. Thus in a recent nationwide survey of rural women in Togo, climatic risks emerged as a main preoccupation. However, mention of this occurred in the context of a list that included lack of labour, delay in supply of improved seeds, cost of transport, sickness, and slump in sales (Togolese Federation of Women 1988: 191–2). More detailed analyses of the impact of drought found that in Nigeria, Hausa women are more vulnerable (Schroeder 1987), as are Bengali women in India (M. Ali 1987).

3 For example, Richards (1986) likens the farming venture undertaken by the Mende of Sierra Leone with a ship, with a crew hired and paid off at each point during the voyage. Each voyage takes place over a catenary profile and through an agricultural calender (i.e., through both space and time), involving the labour of women in some areas, senior women in others, men for certain agricultural activities, and it is only at one particular point in the agricultural calendar that anything approaching a 'farm household' appears at all.

In cases where larger units are significant, such as production brigades in China from the 1960s, the household may not be an appropriate unit for all aspects of analysing access. The household may control some of its consumption, and small plots of land for production, but most resources and the accumulated surplus are outside their control.

4 For simplicity we refer to income opportunities, although a better term is probably livelihood, which implies supporting life without the assumption that this is done through access to a cash 'income'. Livelihoods may include activities

of self-provisioning (subsistence farming, fishing, or pastoralism) in which cash plays an insignificant part.

5 It is the female children who tend to be withdrawn from school when illness with AIDS removes principal wage- or food-earners from Ugandan families (Barnett and Blaikie 1992).

6 'Rules of the game' can shift very rapidly, as in the new regimes that accompany the establishment of colonial rule or the sudden establishment of private owner-ship of land (or, conversely, as a result of the sudden collectivization of land as in the Ukraine in the 1920s and 1930s). O'Keefe and Wisner (1975) show how changes in the 'rules of the game' rendered ineffective a number of indigenous African mechanisms for coping with drought, resulting in increased potential for famine in the colonial period.

7 The other models considered in developing the access model include O'Keefe and Wisner (1975); Wisner (1978b); Sen (1981); Watts (1983b, 1991); Blaikie (1985b, 1989); Winchester (1986, 1992); Swift (1989); and Drèze and Sen (1989).

8 These competing theories of famine causation treated in Chapter 4 include Sen (1983) and Ravallion (1987), who emphasize the behaviour of markets and their impact on the population; Rangasami (1986) and Firth (1959) who deal with the structures of domination and the time–space aspects of disasters; and Hellden (1984) who studies the impact of drought upon famine in Ethiopia.

9 'Ethnoscience' is the term often used for vernacular, local knowledge of the physical environment. Some have used the terms 'people's science' (Wisner, O'Keefe, and Westgate 1977), 'folk science', 'folk ecology' (Richards 1975), 'écologie populaire', 'people's knowledge' (Rau 1991), and 'indigenous knowl-edge' (Brokensha, Warren, and Werner 1980). Within environmental design and architecture the term 'community design' is common (Wisner, Stea, and Kruks 1991). We will use the term 'local knowledge', connoting a broader knowledge base that includes social relations and not just taxonomy, mechanics, chemistry, etc. For a critical review of the use and misuse of local knowledge by outside development agents, see Wisner (1988b: 256–62).

10 Models of the risk-averse farmer abound: see Ellis (1988) for a review.

11 On women's access to resources see Rogers (1980); Dey (1981); Agarwal (1986); Vaughan (1987); Sen and Grown (1987); Carney (1988); Wisner (1988b: 179–86); Shiva (1989); Downs, Kerner, and Reyna (1991); and Schoepf (1992).

12 Examples are given by Cutler (1984); Greenough (1982), writing of the 1943–4 Bengal famine; Vaughan (1987) on Nyasaland in 1949.

13 For South Asia see Agarwal (1990: 367); Fernandes and Menon (1987). On Kenya 1971–6 see Wisner (1980); and Downing, Gitu, and Kamau (1989).

14 Corbett (1988) has reviewed four major studies of coping mechanisms in the face of famine: these are of northern Nigeria 1973–4 (Watts 1983b); Red Sea Province, Sudan 1984 (Cutler 1986); Wollo Province, Ethiopia 1984–5 (Rahmato 1988); and Darfur, Sudan 1984 (de Waal 1987). Brown (1991) presents another detailed account of the coping sequence in Chad, as do O'Brien and Gruenbaum (1991) from two contrasting sites in Sudan. Agarwal (1990) has also reviewed accounts of coping strategies in South Asia.

15 This is true even in more densely populated regions such as South Asia. On common property resources in Asia see Blaikie, Harriss, and Pain (1985); Agarwal (1990); Chambers, Saxena, and Shah (1990).

16 This is especially true at the onset of drought, when it is impossible to know how long reduced or interrupted rainfall will persist and the initial coping strategy is to preserve the basis for continued existence at normal levels afterwards. See Cannon (1991) for a review of these approaches.

17 For other examples of disaster relief, prevention, and mitigation in which verna-
cular coping and innovations from the outside are combined see Wijkman and
Timberlake (1984: 104–43); Timberlake (1985); Harrison (1987); Maskrey (1989);
Anderson and Woodrow (1989); Grainger (1990: 276–321); Harley (1990);
Pradervand (1989); and Rau (1991: 145–205).

Part II

VULNERABILITY AND HAZARD TYPES

4

FAMINE
AND NATURAL HAZARDS

INTRODUCTION

Of all disasters, famine is perhaps the most damaging. There have been more references to its occurrence historically than any other type of disaster, and throughout history the state has been involved with famine far more closely than with earthquakes, floods, tsunamis, storm surges, and other types of disaster.[1] Generally the number of people affected has been far greater in famine, and its social and political impact upon the affairs of state and rulers has been more profound. The worst recorded earthquake disaster caused the deaths of about 240,000 people in 1976 at Tangshan (China), but it is dwarfed by famines that have frequently caused the deaths of more than 2 million. In the Chinese famine in of 1958–61 the death toll is estimated at between 14 and 26 million (Kane 1988) and possibly as high as 40 million (Article 19 1990: 18).

Today, famines still occur. The reasons have undoubtedly changed, but people die in the same appalling ways as in the past. It remains a daily preoccupation of many people, particularly in Sub-Saharan Africa, and in the consciousness of many millions of others who have witnessed famine within living memory (for example, in southern Nigeria, Ethiopia, the Sudan, Bangladesh, and China). It is also a vivid reminder of oppression and imperial callousness, as in the case of the Ukraine under Soviet control in the 1930s (Dando 1980) and Ireland under British rule in the 1840s (Regan 1983).

An enormous literature has grown up to explain why famines occur and what to do about them. However, it is clear today that there is a disjuncture between explanation and policy. Explanation of famine is largely a product of the academic world, and the study of famine has almost become an academic industry. Policies for dealing with famine are a product of famine relief agencies, governments' advisers, and of governments themselves. The lack of affinity between the two types of literature is surprising. In an ideal world there should be a progressive and interacting relationship between the theory of famine avoidance and relief and policy, but instead there tends to be disjuncture. The two sides are separated almost by a different language, and are pervaded by different constraints and concerns. One of the purposes of this chapter is to explain why, and to suggest ways in which they may be reunited.

FAMINES AND CONFLICTING IDEAS ABOUT THEIR CAUSES

Part I of this book questioned the perception that disasters are 'natural' in any simple way. Some preliminary arguments were developed to show that 'social' (in the widest sense to include political, economic, and cultural) reasons have to be given much greater prominence in understanding how disasters are caused. Despite the fact that many government policies and most popular ideas of famine tend to blame drought and other natural hazards for causing famine, we will show why such explanations are inadequate. Part II focuses on the different types of hazards which are normally associated with causing disasters. Famine, although often linked with drought, is a disastrous outcome which can be associated with a number of natural hazards, but also often occurs without any 'trigger' event in nature (for example, as a result of war, or of policy failure, or government policy shifts which alter the basis on which people obtain their food entitlements).[2]

In the case of famine, it is possible that this shift in explanation away from natural causes goes back further than for other disasters. Sen's work (1981) on the Bengal famine of 1943 focused on the purely social causes of famine. The publication of *Natural Disasters: Acts of God or Acts of Man?* (Wijkman and Timberlake 1984) also contributed a survey of much literature on famine that asserted human causation. Wisner (1980, 1988b), Watts (1983b), Bush (1985), Glantz (1987), and many others (particularly the radical writers about the Sahel famines of the 1970s[3]) have argued for the relative unimportance of drought as a cause of famine, except as a 'trigger'. This is an instance of a more general movement to relate environmental processes (e.g. soil erosion, decrease in biodiversity, wild fires, water quality, air pollution, environmental health, etc.) in a more sophisticated way to social processes.[4]

Since the early 1980s, academic literature on the explanation of famine has shifted away from giving prominence to natural events towards an emphasis on social structure and process. Sen (1985: 13) claims that both Marx and Adam Smith directed attention to the social causes of famine, and this suggests that such explanations go beyond a simple divide between the political 'left' and 'right'. As Curtis, Hubbard, and Shepherd (1988: 3) say, 'social, not natural or technological, obstacles stand in the way of modern famine prevention'. They go on to argue that modern famines, more than ever, are due to social processes of population pressure, environmental degradation (which we would claim is a socially-determined process), decline in self-provisioning, and the reduction of alternative means of earning or producing enough to eat. Any explanation of these social pressures must be ideological, and it is difficult to talk about the relationship between drought, flood, frost, climatic change, or environmental degradation and famine without becoming involved in statements with a high degree of political content.

One of the main sources of confusion on the subject arises from the multiple causes of famine and their great variety in space and time. Clearly a rigorous and tightly conceived approach is required. Some of the literature makes a distinction between 'general and predetermining' factors and the 'trigger' mechanisms of the hazard impact. Torry (1986) uses similarly distinct 'ultimate' and 'proximate' causes of famine. In the language of the models discussed in Part I these would be termed the *root causes and underlying pressures* which create *unsafe conditions*. When these conditions prevail, a famine may result from natural events such as drought, flood, and pest attack, which can thus be perceived as the 'trigger' of famine, but not its ultimate cause.

There are numerous explanations of famine, and it is useful to distinguish between 'general and predetermining' causes (or root causes in our PAR model) and specific and more focused causes which channel root causes into unsafe conditions. These latter take effect at the 'point of pressure' in the PAR model, and involve detailed analysis of the unsafe conditions and how they affect the outcome of the disaster when and as it occurs. The specific causes at the 'point of pressure' require much more rigorous analysis because they must describe why specific people, at a specific place and time, cannot eat enough food to survive. They must show accurately the movement of food to people over time, and also account for the reasons why some cannot avail themselves of that food.

As this chapter shows, there are two main (and largely competing) types of explanation of famine, based on differing sets of causal mechanisms. The first concentrates on famine as a result of food availability decline (often abbreviated to FAD). The mechanism is essentially simple and common sense. Any number of factors may be invoked, such as natural events like drought causing crop failure (or reduced growth of pasture for livestock or both). This in turn reduces the aggregate amount of food available, so that famine is seen simply in terms of there not being enough food to go round.

The alternative mechanism involves the decline of some people's entitlement to food (abbreviated to food entitlement decline, or FED). According to this theory, famine is a result of the (many and complex) ways in which people's access to food is reduced because of the operation of social or political processes that deny or lessen their 'entitlement' to food. These may involve a deterioration in the ability of people to grow their own food or buy it through various forms of exchange (especially through sale of their labour). To this context should be added the impact of various natural hazards which may not reduce the overall amount of food, but instead affect the success of different groups of people in fulfilling their entitlements. For instance, in drought conditions market prices of food normally rise often as a result of processes that are independent of how much food is really available. Self-provisioning food producers may experience lower harvests, but some may have stores to fall back on. Others who rely on wage-labour, petty trade or artisanal production do not get enough to eat, because their cash income no longer matches increased prices. If such conditions continue over a long enough period, then some of them will die. There may well be enough food in aggregate to prevent famine, but some do not have the physical, social, or monetary means to avail themselves of it.

This sort of explanation focuses much more firmly on relations of power within a society which may account for the distribution of assets and income (unequal in 'normal' times) which become a matter of life and death in times of famine. Using this explanatory model rather than FAD tends to reduce the causal importance of natural events which, although they may be linked to a decline in the aggregate supply of food (with the impact of drought, flood, or pest attack), are analysed in the context of the political economy of root causes and predetermining factors. In other words, people are made vulnerable to the impact of a natural hazard by their place in the economic, political, and social processes that affect their exchange entitlements.

An explanation of famine requires that both root causes/underlying pressures and the processes that generate unsafe conditions (e.g. FAD or FED) are linked together in a satisfactory causal chain. Therefore the conflicting explanations involved in FAD and FED need to be resolved so that it can be shown how the root causes are channelled through well-defined and precise mechanisms towards causing a famine. Without a resolution of this conflict, explanations remain speculative and conditional and root causes of famine are not firmly attached to actual famines. For example, the 1973–7 famine in central/north Ethiopia has often been attributed to drought. But such an explanation begs the question of why, in Ethiopia, there have been droughts without famine, and famines without droughts (for the historical record, see Pankhurst 1974; Kapuscinski 1983).

We need to understand the complex linkages that lead from root causes to the unsafe conditions that can make people vulnerable to famine. Such vulnerability may expose them to a number of trigger events (of which drought is only one). In the 1970s case, below average rainfall had been recorded in the area (Asefa 1986: 19), but there is still much debate whether and how this was a direct factor in the famine. Was there export of food from the area at the time of famine? Was the decline in aggregate food supply alone enough to precipitate the famine? These issues can only be debated in terms of the precise mechanisms by which the famine was caused, and are examined in more detail in the section on access later in this chapter.

To take another example, such global dynamic pressures as war and environmental degradation have frequently been identified as causal factors in famine.[5] In all but a few cases, however, these attributions have more of an ideological than an analytical purpose. Knowledge of the precise mechanisms that link war or environmental degradation to actual cases of famine remains inadequate.

An Oxfam publicity booklet (Cater 1986: 1) illustrates the problem of explaining the famine in the Sudan, proposing the following reasons for famine: 'The people who died did not do so because the rains failed in 1984. Despite hard work and calculation, they died because they could not grow enough food and were too poor to buy what they needed.' There follows a brief description of some of the factors which combined to bring about a situation of 'not enough food' and 'too poor to buy'. These included drought, unsuitable technologies for providing water for humans and livestock, population pressure and a fragile ecosystem, deforestation and a fuel crisis, chronic uncertainties over land tenure, lack of credit, monopolistic and monopsonistic power of merchants in rural areas.

This booklet and other more academic literature (e.g. Shepherd 1988; de Waal 1987) provide good evidence of the importance of root causes and pressures in the case of the Sudan. Curtis, Hubbard, and Shepherd (1988) add low income, poor import capacity, and war disruption to Oxfam's list. Walker (1989) cites the overcultivation of fragile sandy soils (*qoz*), drought, loss of access to land through expropriation by absentee merchants and landlords, widespread reliance on unreliable waged employment and thus on the market to buy food in situations of sky-rocketing prices.

Thus there is a measure of agreement about causes in the case of the Sudan. But the task remains to explain how these factors combined at a particular place and time to prevent many people in Kordofan and Darfur Provinces in the Sudan from eating enough food. Most of these factors also apply to the famine that emerged in southern Sudan in 1988, but in different combination and process. From this we may conclude that none of these predetermining factors are either necessary or sufficient to cause a famine, and explanations must rely upon precise mechanisms (including FAD or FED or a combination of these models) which bring into effect different combinations of the various root causes and dynamic pressures.

This brings us to the second point about the explanation of famine. If there are many combinations of factors and mechanisms which bring about famine, then each famine is unique. The task of building theories of famines is particularly difficult because of the complexity of each specific case. It will probably involve not only an understanding of the existing system of production, but also the distribution of food in terms of access to land and inputs as well as the operation of the market, the determination of prices, and the behaviour of traders in food staples (Cannon 1991). Government policies with regard to food production and distribution and in famine relief itself may also have a profound effect. Then there is always a series of contextual events peculiar to each famine, a 'sequence of events' (Alamgir 1981), or in Currey's term (1984) a 'concatenation'.

Therefore the narrative of each event will be an important element in the explanation of particular famines, and at all times it is advisable to maintain a flexible analytical approach. For example, *Silent Violence* (Watts 1983b) is a book of over 600 pages which traces the changing contexts of famine in northern Nigeria. His research design is like a set of Chinese boxes or Russian dolls in which the local level of explanation is dependent on an understanding of the regional, national, and international levels. Each instance of famine therefore is a 'concatenation', but they are related to each other by continuous changes in social relations of production, climatic fluctuations, and the like. In this way the impacts of changes in the political economy, or of climatic change, are linked to larger-scale and broader processes which can be characterized as root causes and underlying conditions, in the way that the chain of explanation in the PAR model has demonstrated.[6]

REVIEW OF EXPLANATIONS OF FAMINE

Indian Famine Codes

Perhaps the first coherently written explanation of famine, linked to policy recommendations, is the Indian Famine Commission Reports and Famine Codes. These date from 1880 and were used by the British until Independence in 1947 (and in modified form since). The Reports contain many speculations about the causes of famine, and almost all the major types of explanation are put forward. Drought and crop failure are suggested both as direct and indirect causes. The operation of the market, trader behaviour, the (usually favourable) impact of the British-built railways, and the whole issue of the price of food staples are all mentioned at different times.

There is a very strong ideological stance derived from the work of Adam Smith and John Stuart Mill, with a professed dislike of interference in the operation of the market through price controls, and the belief that free trade is the best guarantee of satisfying effective demand. There is also an aversion to charity and free hand-outs to victims and a strong ethic of 'self-help' runs through the Reports, which is reflected in the policy recommendations in the Codes.

The backbone of famine relief was massive public works generating guaranteed employment and free assistance for those unable to work. That such interventions breached the ideological opposition to interference in markets was perhaps a contradiction essential for those who wanted to prevent people dying. Four 'tests' were established to ensure that only the deserving received relief. There were detailed instructions in the Codes about early warning signs of impending famine, the duties of the police, medical officers, and other local officials, wages and rations, famine relief works, and many other practical instructions.

The effectiveness of the Codes is still the subject of heated debate, with rather an overstated radical and nationalist critique at one extreme, and an imperial apologist defence at the other. Drèze (1988) and McAlpin (1983) both make measured claims for the effectiveness of the Codes in preventing famines, although serious exceptions were admitted in 1896 7 and 1899–1900. They have formed the basis of post-Independence famine prevention and relief. For example, the Maharashtra drought of 1970–3 was effectively prevented from triggering a famine by an employment guarantee scheme similar to that provided in the Codes. Similar policies formed the basis of famine regulation elsewhere in the British Empire (for Nyasaland, now Malawi, see Vaughan 1987; on others Curtis, Hubbard, and Shepherd 1988: 39).

Three characteristics of the Reports and Codes are outstanding and for several reasons are particularly instructive of the relation between the theories of famine causation and policies of prevention, relief, and rehabilitation. Firstly, there was a clear recognition of most of the 'root causes' and of the 'pressures' causing famines. Secondly, ideological preconceptions influenced this recognition and had a very significant role in policy. Thirdly, the Reports do not involve a single theory or explanation of famine, but a number of loosely articulated but well-argued observations. From these and not from a single dominant theory, reasonably effective policies were formulated which have stood the test of time. Indeed, some commentators on contemporary theoretical debates and policy formulation think that the Indian Famine Codes covered most of the ground over 100 years ago.

Influence of disaster vulnerability analysis

Despite this, much of the writing about famines during the twentieth century continues to emphasize the importance of natural events in explaining disasters. While a few writers emphasized the social causes of disasters (e.g. Gini and De Castro 1928), the dominant explanation emphasized the natural causes of disasters including famine.[7]

It took until 1983 for a major coherent critique of the notion that disasters are explained by reference to natural factors to appear: in the form of *Interpretations of Calamity* edited by Hewitt (1983a). The main thrust of the critique was that if disasters, particularly famines, were attributed to natural causes, they could be explained in terms of exceptional events and not of continuing and normal social processes. Famines could be attributed to *un*precedented, *un*natural, and *un*expected events, and therefore appear to be quite separate from 'normal' life. In this framework, sudden-onset natural phenomena, or even slow-onset drought, could provide the explanation because they constituted the new, decisive, and therefore principal cause. In

Hewitt's view (1983b: 9–24), the explanation of disasters should rest more fully on a social analysis of the processes which create the conditions under which 'exceptional' natural events triggered disasters.

This analysis focused on the social processes of impoverishment and exploitation which expose people to hazards (make them vulnerable) as a normal and continuing part of life. An explanation of disasters in terms of natural events invites technological solutions (rather than social ones) to the containment of floods, design of earthquake resistant buildings, and the introduction of a more productive technology in agriculture. When this approach demonstrably fails (as it sometimes does), the problem is perceived to be beyond the powers of technology and is once again thrown back into the lap of 'exceptional' and 'unprecedented' events.

Hewitt (1983b: 12–14) argues that a previous generation of academics and practitioners virtually ostracized those who sought explanations of disaster that went deeper than the impact of the natural hazard. Given the dominance of science and technology in the modern era, the authors of any analysis of causes that did not suggest that hazards could be modified and respond to technology (eg. satellite early warning systems and reinforced concrete) were exiled from mainstream social explanations.

Natural hazards and link with food availability decline

This chapter suggests that the role of natural events and social processes with specific reference to famine should be examined more closely. However, this does not mean that we exclude natural events completely from the explanation of famine. Social processes themselves are sensitive to the impact of natural forces that alter people's entitlements to food. Instead, the approach used here is to treat natural events as direct causes of famine, but to consider their impact as entirely dependent upon social mechanisms which determine who has a sufficiency of food and who does not. Thus all natural events (flood, drought, or pest attack, for example) fail either as necessary or sufficient causes of famine.

This is not to deny that they may have a powerful general impact in both the short and long term. Detailed study of those natural processes (e.g. moisture stress in crops, climatic change, or rainfall statistics) is important and helpful. Considerable progress has been made in these applied scientific areas in recent years, and it is timely and relevant.[8] The remaining challenge is to combine the results of such work with the growing body of administrative experience and recent insights on disaster preconditions and human vulnerability.[9] As noted earlier in this chapter, there is still a gap separating people engaged in these types of analysis and applied work.

The school of thought which attributes famine to an aggregate decline in the food supply (FAD) is clearly linked to explanations of famine in terms of natural events, particularly drought. To identify drought as the immediate cause of crop failure and therefore of a decline in food supply seems a common-sense deduction. In recent literature it is difficult to find pure supply-side explanations of famine. But in the balance of emphasis given to a decline in aggregate food supplies and its immediate causes (including drought) on the one side, and the detailed mechanisms which actually precipitate a famine on the other, the emphasis is often on the former.[10]

The major problems with a simple causal connection between a fall in aggregate food supply and a famine are the assumptions that food availability is shared equitably among the population, and that its members have no other source of income than food production. Both are usually unwarranted. While many explanations of famines start with a trigger event such as flood or drought, there are usually difficult anomalies between the relative severity of the famine and the decline in food supply in space and time. For example, Currey (1981) shows by mapping local-level data that famine deaths in Bangladesh do not closely relate to production of food staples at quantities below the average. Likewise, Kumar (1987) shows that the number of districts reporting normal, above normal, and below normal aggregate crop production for 1972–3 in Ethiopia does not correspond well with the spatial distribution of famine-related mortality. He also states that the evidence points to a fall in food production *after* the main impact of the famine had been felt (Kumar 1987: 13). Similar difficulties confront almost every case of famine for which reasonable production and mortality data exist.

The conclusion to be drawn is that natural events are implicated as a trigger to famine, and falling aggregate food supply either in the short or medium term usually makes an area more prone to famine. Mellor and Gavian (1987) argue that the long-term problem of abysmally low agricultural production in Africa remains crucial to explaining the risk of famine. However, Wisner (1988b: 148–86) presents evidence from Kenya to show that a policy of *national* food security can actually undermine *family* food security if efforts to achieve the national objective mean promoting the interests of a minority of wealthy farmers at the expense of support for the majority of smallholders. We must also keep in mind that famine may occur with no natural trigger involved at all, and there are arguments about areas experiencing famine supposedly caused by drought (Ethiopia) or disease (the Irish famine) from which exports of foodstuffs continued.[11] Such situations provide support for our emphasis in this book on 'sustainable livelihoods' and access to resources and entitlements.

A recent analysis of famine in Ethiopia by Diriba (1991) has amply shown how long-term secular declines in cereal production per household and per hectare (as a result of population growth, land shortage, and environmental degradation) have made self-provisioning very difficult. Even in 'normal years', most households have to buy food on the market to make up for shortfalls in household production. However, onerous taxation by the state (a priority claimant on income) and the lack of alternative income-earning opportunities make this entry into the market very risky for the majority. Thus a decline in aggregate food supply within an area is an important factor, but, as always, the lack of alternative means to purchase food turns this long-term decline into a potentially disastrous situation virtually every year. Goyder and Goyder (1988) and Kebbede (1992) agree that excessive taxation by the Ethiopian state has tended to push many marginal peasant producers over the edge, when combined with other adverse conditions.

Elsewhere in Africa the predations of local administrators and warlords have discouraged people from farming. In eastern Zaire men took to illegal gold-mining and smuggling as a way of feeding their families (Newbury 1986); while in Mozambique destabilization by armed bandits has forced many millions to abandon their fields (D'Souza 1988; S. Smith 1990).

Markets and market-failure explanations

Another set of explanatory mechanisms of famine concerns the operation of markets, which in times of hunger may be unable to meet the demand for food of a set of people, even though that effective demand (backed by the ability to purchase) exists (Seaman and Holt 1980). In other words, the market response which should supply the food that is demanded is unable to operate properly. Poor transport is often suggested as one of the major reasons for this failure. Emerging from this is a related argument that modern transport can reduce famine mortality. Conversely, a number of radical writers (reviewed in the next section) attribute a dangerous decline in self-provisioning to the spread of commercialization, which is often aided by expansion of railway and road systems.

Drèze considers the Indian situation under British rule in relation to the expansion of railways:

> Of the fact that the expansion of the railways resulted in a greater tendency towards uniformity of prices there can be little doubt . . . .
> One may also generally expect a reduction of price disparities to reflect greater food movements towards famine-affected areas, and to result in an improvement of the food entitlements of vulnerable sections of the population in these regions. However, it is easy to think of counter-examples of which two are particularly important here.
>
> (1988: 19)

The first of these counter-examples is that price equalization also went on at the international level, with large-scale export of grain abroad during famine periods. The second is that the equalization of prices brought about by the introduction of railways is not the same as the equalization of real wages or food entitlements. Thus there could be an area which exported grain but which was also a low wage area, with the result that labourers (unable to grow food themselves) would be unable to afford to purchase the grain. Strictly speaking this is not a market failure, but an example of how an efficient, price-equalizing market could precipitate a famine.

Other factors may impede the transfer of food from surplus to deficit areas through the attraction of higher prices in the latter. It can be argued that inadequate integration of markets can constitute another constraint on perfect market operation. For example, there is E. Clay's argument:

> If Bangladesh is regarded as a fully integrated production system with smooth inter-district flows of commodities, then there is no production problem. But to the extent that there are frictions and difficulties in moving commodities between districts, regional losses of production can have severe effects on food prices, intensifying the effects of loss of production, income and employment.
>
> (1985: 203)

Thus the problem is seen not as a fall in aggregate regional food production but failures of the market which, if perfect, should regulate imports and satisfy effective demand. Also expectations of future food shortages can lead to speculation and hoarding. Indeed hoarding may be desirable given the expectation of future scarcity. In some cases a lack of information about demand may also prevent the market mechanism from satisfying effective demand. For example, Ravallion (1985) found some evidence for a systematic bias in rice prices during the 1972–5 period in Bangladesh, caused by inaccurate and pessimistic information about food prospects in the press. This led both traders and better-off consumers to hoard, so causing further price rises. Therefore according to this view, it is not a decline in food production which causes famine (it could be imported) but the fact that the market has failed to receive the correct signals to distribute food where it is demanded.

Failures of entitlements

In 1981 Amartya Sen brought together a number of disparate sources to significantly advance the debate on famines. It centred on entitlement failure, or the notion of food entitlement decline (FED) which has been briefly discussed above. It is the failure of effective demand (or 'pull failure') which can cause famines, rather than the market failure (or 'response failure' as Sen (1985) terms it) suggested by other theories. People may need to purchase food but do not have the necessary cash or other exchangeable resources to purchase it. Therefore, in a pure case of 'pull failure', effective demand is absent and does not maintain an upward pressure upon prices. The approach distinguishes between aggregate availability or supply of food and an individual's access to, or ownership of, food. People obtain food through five different types of 'entitlement relationships' in private-ownership market economies. These are adapted from Sen (1981: 2) and Drèze and Sen (1989: 10):

1 Production-based entitlement which is the right to own the food that one produces with one's own or hired resources.
2 Trade-based entitlement which describes the rights associated with ownership when they are transferred through commodity exchange.
3 Own labour entitlement which is the trade-based and production-based entitlements when one sells one's own labour power.
4 Inheritance and transfer entitlement which is the right to own what is given by others (e.g. gifts) and transfers by the State such as pensions.
5 Extended entitlements which are entitlements which exist outside legal rights (e.g. of ownership) and are based on legitimacy and expectations of access to resources.

In our access model, these entitlements are conceived of as socially-derived 'resources' which can be called upon by individuals according to the current allocative 'rules of the society'. These include rights and obligations between patron and client, kin, husband and wife, mother and father and child, elder and junior, or beggar and alms-giver. Two further entitlements in the later work of Drèze and Sen (1989) make Sen's model approximate to our access model. It is worth commenting that the latter does not claim to be a theory, but rather a structural approach and a structured checklist of factors affecting access. Sen's model has therefore also become an all-embracing theory, and as such no longer competes with other theories of famine such as FAD.

Entitlements are either owned by a person (described by Sen as the person's 'endowment') or can be exchanged by that person for other commodities (termed 'exchange entitlement'). People are vulnerable to starvation if their endowment does not contain adequate food or resources to produce food, and their capacity to exchange labour or other goods and services cannot be translated into enough food. This can occur without a decline in aggregate food supply and without any disruption or malfunction of the market. Such events were possible despite the equalization of the price of food staples brought about by the introduction of railways in India.

With shifts in entitlements a person can suffer either a fall in endowments (such as crop failure or livestock deaths), or a fall in exchange entitlements (in which food prices rise, wages fall, or demand for one's own production falls so that the terms of trade with the market for food shifts unfavourably against the individual), or a combination of these. Analysing exchange entitlements as an additional mechanism for failing to obtain enough food permits explanation of famine occurring in boom conditions, as in the Irish famine of 1846–7, or the Ukraine famine of 1930 (during which there were forcible exports of grain to the rest of the Soviet Union which caused widespread famine in the republic).

The analysis of food entitlements decline (FED) was undoubtedly a great advance on theories of food availability decline (FAD) for a number of reasons. Firstly, FED acknowledges the importance of changes in purchasing power. Secondly, it disaggregates regional food production and availability, and follows through how food is distributed to individuals. It permits analysis of intra-household food allocation, though this is less well studied (Shepherd 1988). It explains why the rich never die in a famine and shows, indeed, how some classes become wealthy during a famine while others die. Thirdly, it involves the regional, national, and world economy in the analysis and draws attention to the possible prevention of famines by food imports. In general terms it is an active model – governments *can* intervene to rescue endowment or entitlement failures, but are also scrutinised in terms of their contribution to causing the problem in the first place.

Food entitlements decline (FED) also exposes the shortcomings of the food availability decline hypothesis (FAD). Firstly, FAD only deals with supply factors while FED claims to deal with both. Therefore the latter can address the impact of rising prices of food staples. Secondly, FAD cannot deal with disaggregated populations and explain why some starve while their neighbours do not. This point is well-illustrated by a story from the 1970s Sahel famine recorded by Mamdani (1985). A fat man is reported to have said to a thin man 'You should be ashamed of yourself. If someone visiting the country saw you before anyone else, he would think there was a famine here.' Replied the thin man, 'And if he saw you next, he would know the reason for the famine!' Thirdly, the FAD hypothesis is ill-equipped to identify the social causes of vulnerability and poverty other than in general terms of factors such as low agricultural productivity or backward technology, which do not lead to analysis of underlying social determinants.

Criticisms of the entitlements approach

On the other hand, there have also been criticisms of FED, and particularly of the way in which Sen has sought to defend it. Firstly, there is the scale and boundary problem. If the analysis is stretched to include a big enough area, of course there is enough food to avert a famine in part of it. Secondly, some famines clearly have had their origins in FAD, even if the government of the day engineered it or neglected the warnings. The debate between Sen and Alamgir concerning the causes of the Wollo and Tigray famine of 1973–4 in Ethiopia focused on food prices. Did they rise and was it in the context of effective demand? Or did they not appreciably increase, in which case there would be support for Sen's 'pull failure' hypothesis? The argument here revolved around 'price ripples' where effective demand was not met by supply because of high transportation costs and either a lack of information reaching traders, or their unwillingness to risk distributing small lots of grain to scattered markets. These market imperfections in the presence of effective demand caused outmigration in search of food and price rises in a series of ripples resulted.

Clearly, many famines have been preceded by FAD, and although it may be incorrect to identify FAD as an ultimate or even most important cause, it is inescapable that a fall in the amount of locally produced food (because of war, drought, or longer-term environmental decline) calls into question the ability of people to find alternative sources of food. Cutler (1984) and Devereux and Hay (1986) point to FAD being a major mechanism of famine in Tigray and Wollo where the market was not well developed, transportation very poor and expensive, and the proportion of food bought or sold in the household very small. This type of explanation is also borne out by Diriba (1991) in his study of an area in central–south Ethiopia.

Initially Sen's notion of FED tended to perceive endowments and entitlements as static and given. This weakness is addressed in more recent developments of these ideas. In fact they are fought over (Watts 1991), and constitute the terrain of struggle within societies in which group interests (defined by class, caste, gender, age, and ethnicity) are in contradiction.

In conclusion, the FED model (the entitlements approach to analysis of famine) has released famine study from theoretical constraints and played a politically progressive role. However, the FED–FAD debate has been polarized and increasingly couched in the style of an academic contest. This pursuit of a single theory of the mechanism of famine has diverted attention from multiple causality and the possibility of famines at different times in the same place being caused by a different mix of factors. What is needed now is a further development of the concept of vulnerability which FED mentions but does not develop in detail (e.g. income, assets, class, occupation, bargaining power, and exploitation). This is provided by the development of an access model of famine in the next section.

ACCESS AND FAMINE

A more specific and detailed development of the generalized access model described in the last chapter is shown in Figure 4.1. In box 1 the broad political economy (social relations and flows of surplus) provides a context in which individuals and households earn a living. At the international level, certain aspects of the world economy impinge in an indirect manner on food consumption. For example, national foreign exchange reserves affect a country's ability to import grain to avert possible famines. The long-term decline in world food stocks from the 1970s, combined with the entry of the then Soviet Union in the world market, has affected the ability of some countries with low foreign exchange reserves to buy in or hold adequate stocks of grain. At the national level the nature of the state itself is extremely important in explaining in general terms the ability or the willingness to implement policies that reduce famine or lessen its impact.

Box 1 also provides a caricature of a political economy of civil society, focusing on the social relations of production of the peasant household. Other classes such as merchants, moneylenders, and landlords are usually important groups in a famine, although they seldom suffer the effects of it, and on the contrary sometimes make a fortune. The social relations of production are important basic determinants of the distribution of access to the means of growing food or the income with which to buy it.

A number of radical critiques of technocratic and *status quo* explanations of the Sahel famines of the 1970s have been developed largely from this context (see note 3). In a similar vein, Shindo (1985) and Mamdani (1985) link state military expenditure and exploitation of peasants (through unequal market relations and coercion) to the development of famine. Some of the assumptions of this literature are perhaps naive or wishful thinking. A reduction in military expenditure does not automatically mean that the saved funds or lower taxes are channelled into reducing vulnerability and famine proneness. Unfortunately, most of the radical literature is also content to explain famine in terms of root causes and does not indicate specific mechan isms. Our access model specifies such mechanisms.

The individuals in households (box 2) are therefore put in the context of the wider society and their assets and resources are listed here. The physical assets (equivalent to Sen's endowments) usually include access to land, seed, agricultural implements, etc. Other assets are social and refer to the membership of a tribe, segmented group, or family as well as to rights to assistance in the form of loans, employment, food, etc. As such they are defined more widely than Sen's entitlements, and include various forms of expectations (or 'claims' in Shepherd (1988)) and the ability to mobilize resources commanded by others, such as labour and food in the form of gifts.

In box 2, the first column shows men, women, and children in pictorial form. It is important to distinguish between people of different gender and status within the household – something which Sen has taken up in later writings (1988, 1990). Sometimes 'older sisters will serve food to younger brothers . . . and female etiquette demands that women take less food and eat more slowly in the presence of the men and of guests' (Rahmato 1988: 237). Some survival strategies prioritize the food intake (and ultimately survival) of men over women and adults over children and old people. It is important to establish whether these habits persist in times of actual famine. According to Rahmato they do not, but in other accounts there is considerable struggle between men and women over food resources during times of stress (Goheen 1991; Kerner and Cook 1991; Schoepf and Schoepf 1990).

Each of the people represented pictorially consumes a certain amount and kind of food during 'normal' times. This provides nutrients which are absorbed and utilized by the body to the degree allowed by that person's state of health. Thus health and nutrition interact, and the resulting baseline nutritional status influences the ability of that person to survive food and

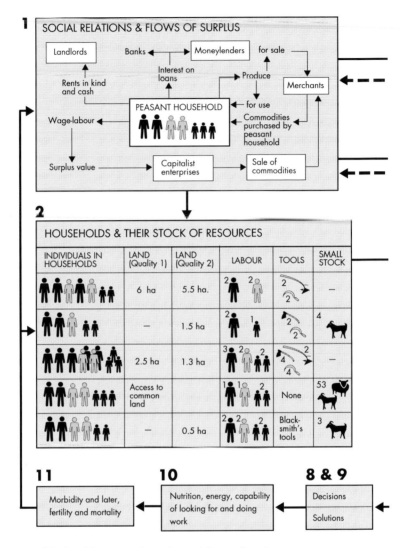

Key: black adults are males; white adults are females

Figure 4.1 Access to resources to maintain livelihoods: the impact of famine

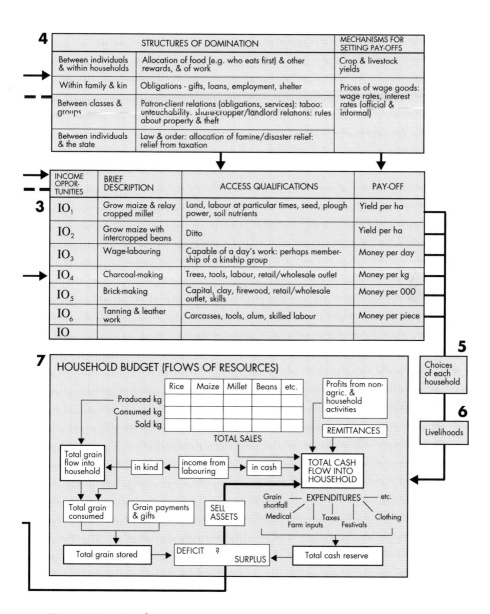

Figure 4.1 continued

health emergencies (see Chapter 5). 'Normal' food intake also influences the working capacity and productivity of household members as they utilize available resources.

Each household shown in box 2 has a range of resources and assets at a point in time. The household then reviews a range of income opportunities, each with its access qualifications and pay-offs as described in the general access model (box 3). This outcome reflects the contested interests within the household (between women and men, perhaps between generations, and siblings). The idea of the household possessing a 'joint utility function' has recently been under attack. The pay-off is a function of the relations of production, the technology used in production, rainfall, fertility of the soil, etc.

Box 4 shows structures of domination and allocation of resources, and is the economic, social, and political expressions of class and gender relations described in box 1. These structures operate at the individual level within the household (usually revolving around gender and age but also genealogy). They involve relations within family and kin, between classes and groups such as patron–client relations, plus the gender division of labour. They also affect expectations regarding crops and property, particularly those safe-guarding private property (or sanctioning theft under abnormal conditions). They also operate anonymously in terms of the market, although there may be an important spatial patterning of its impact upon individuals. Finally, the state provides a broad framework in which rural households have to pay taxes, are subject to the operation of law and order (which may be exercised in favour of particular groups against others), and maybe there are benefi-ciaries of famine relief. The structures of domination and allocation in a specific case have a major part to play in setting pay-offs for different income opportunities and determining their access qualifications.

Each household holds a particular package of endowments and entitle-ments (which it must also be able to store, both physically as grain and as rights and expectations for future realization). This determines the state of the household budget, at a point in time outlined in box 7. Here it is important to specify different foodstuffs and cash as it flows through the household budget. This is a familiar model of a peasant economy in normal times.

In order to trace the triggering of a famine, it is also necessary to iterate the model (showing the succession of annual cyles), but with the system put under stress from a variety of causes. For example, a war can act as a trigger. Leftwich and Harvie (1986) point to the disruptive effects of war on the production and distribution of food. Ethiopia (1984 and 1990), Angola, Chad, the Sudan, Mozambique (1984), and Bangladesh (1974) are all well-documented examples. War tends to reduce both the area sown and labour allocation to field preparation on account of a shortage of labour. Also, occupying armies commandeer food and deliberately burn crops in rebel-held areas. Normal trading patterns are also disrupted so that both food availability decline and entitlement decline can occur. Military authorities often refuse safe passage for relief workers to areas where hunger or even famine serve to put pressure on their opponents (Minear 1991; Messer 1991; S. Smith 1990; UNICEF 1989).

Drought, flood, and biological hazards such as pest attack are other more common triggers (e.g. the Sahel 1971–6 and Darfur 1985, different parts of Ethiopia in 1973, 1984, and 1990). The so-called 'potato blight' (*phytophtora infestans*) provided the trigger in Ireland in September 1845, although the political–economic root causes and underlying factors had been developing through a long period in the history of British colonialism in Ireland (Regan 1983). This example is taken up again in Chapter 5. Kjekshus (1977) details the devastating effect of the livestock disease rinderpest on East African herds in the 1890s and its role as a trigger for famine.

For whatever reason (in terms of a trigger event) a period of food shortages starts. The components in boxes 1–6, which had hitherto been changing relatively slowly over a matter of years or even decades, start to change rapidly. Survival strategies are put into action which mobilize new resources. The household budget for those with a less well-developed access profile suffers, and food stocks in some households run down. The operation of the market also becomes crucial when producers have to enter it to support themselves, and those without access to the means of producing food have to purchase it. The economic debates about the operation of the market for food grains outlined above (in the subsection on markets) are relevant here. At the same time the 'anthropology of famine' is particularly significant and has been underestimated in many accounts (as argued by Rangasami 1986). The important issue is the degree to which the rules and institutions of society can adapt in abnormal times.

For some authors, societal breakdown is both a marker and cause of famine (Alamgir 1980; Currey 1978; Cutler 1985). In the following example, however, Rangasami describes Firth's (1959) analysis of the Tikopia famine in Polynesia, arguing that there was no major 'societal breakdown':

> The skeleton of the social order was preserved though attenuated in content. He [Firth] offers interesting insights into famine: what the famine did was to reveal the solidarity of the elementary family. Even at the height of the famine, it appeared that within the elementary family full sharing of food continued to be the norm.
>
> (Rangasami 1986: 1597)

Boxes 10 and 11 refer to changes in the capabilities of the individual and household to continue effectively to cope with stress, and avail themselves of food by whatever means. If food intake falls for long enough, the capability of people to look for and perform work becomes impaired. If food shortages persist, morbidity and later mortality increases. The cycle is completed by the arrow leading from box 11 to boxes 2 and 1. After some weeks of a famine, the assets and resources of a household may be seriously impaired. Assets may have been disposed of to purchase food (cattle sold or land mortgaged or seed-corn eaten). The members themselves, whose labour is an asset, may be sick or have died.

The arrow also leads back to box 1 because class relations and the associated structures of domination and allocation (box 4) may also shift rapidly as a result of famine. While some households suffer, some merchants and moneylenders amass fortunes by selling grain at inflated prices, foreclosing on mortgages (especially for land) on the non-repayment of debts, and may also gain bonded labour in return for food. However, not all famines benefit the dominant classes, and they can be the catalyst for sudden political change, as, for example, Ethiopia during the last years of the reign of Haile Selassie. They may also indelibly alter a country, as with Ireland after 1845–8.

This approach to famine analyses its structures and processes in relation to making a living in normal times. It is an iterative approach in which 'external' shocks and triggers (such as drought or war) have their impact upon the structures and processes of political economy. With the impact of new external factors, the political economy is subject to change which affects individuals and households differently. Most current theories of famine can be located within different parts of this model and cross-disciplinary insights be gained. In a more schematic manner, our 'pressure and release' model of famine summarizes the same relationships, and links them to global pressures on the one side and to the natural trigger events on the other.

POLICY

Policy options can be examined in relation to the 'chain of explanation' which accounts for famine. This chain involves analysing the 'mechanisms' which produce vulnerability by tracing back from the immediate and local causes (in space and time) to the less direct, longer-term, and structural reasons (underlying dynamic factors and root causes). The chain of explanation runs parallel with the PAR model presented in Chapter 2, in which the root causes are linked to the more direct pressures that create vulnerability. When these coincide with adverse local conditions, the impact of a hazard event may trigger a disaster.

A key issue is to address the policy actions to the level of problem which can effectively be altered. While an integrated explanation of a disaster may be intellectually fulfilling, a policy has to locate itself at a level at which it can make a significant impact. There must be effectiveness in the short run – human lives often depend on it. Yet there is also the opposite pitfall awaiting the policy-maker. As R.C. Kent warns:

By dealing with disasters as isolated phenomena, we lose a sense of the real causes of vulnerability. Conceptually, it is a way of avoiding the full implications of the causes and solutions to disasters. One need not address the global inter-relationships between international trade, currency fluctuations, geo-political and commercial interests and a flooded delta in Bangladesh. In practical terms, by isolating disaster phenomena, one can demonstrate goodwill and test one's technological solutions without being 'mousetrapped' into more long term commitments.

(1987a: 174–5)

For each link in the explanation of a famine there are a range of policy measures. At the international and national level, achieving aggregate food security has been an important policy goal, particularly since worldwide global stocks of grain declined and became much more unstable from the mid-1970s.

In a world with shrinking and less stable trade surpluses of staple grains, national food self-reliance is frequently seen as a defence against famine. However, it is very far from being a straightforward or sufficient one (Wisner 1988b: 158–86), and there have been major famines in Africa and Asia in the last two decades despite there being large grain surpluses in North America and the European Community. In identifying the most famine-prone countries in Africa, Shepherd (1988) considered that those which had suffered the most severe long-term decline in food production, and had the lowest foreign exchange reserves per head, were also among those which had suffered the most severe famines. Yet although inability to purchase on the world market is a dangerous situation for such a country, declining output at national level means little when interpreting shifts in access and the level of vulnerability of different areas and particular groups of people.

The theoretical advance which Sen's entitlement theory afforded has made it necessary to re-evaluate the importance of food security at the national and international level, and refocuses it as 'entitlement security' for the household and individual (Cannon 1991). However, national food security would seem to remain an important consideration in famine prevention, provided it is linked with others. The experience of Ethiopia has been particularly instructive because some claimed that food stocks were used to feed people other than the hungry from Eritrea and Tigray. Also, the 'collective self-reliance' achieved to a modest degree by the southern African countries forming SADCC[12] has enabled exchange of food grains among members. This lessened the impact of FAD as an element in famine potential during the 1980s. In 1992 the whole southern African region faced the worst drought in the twentieth century, and countries such as South Africa and Zimbabwe that normally export to others in the region needed to import food themselves. Ultimately, then, at the level of large regions, concerns over FAD remain important.

The insights provided by FED suggest that a more disaggregated approach to food security is also required. It is necessary to 'map' the vulnerability of different groups as R.C. Kent terms it (1987a: 179), and to do this an understanding of the local economy is required. The access model provides an approach to achieving this. The distinction between an FAD and an FED mechanism is important at this disaggregated scale. An FAD explanation suggests that the best treatment for famine is to reverse the decline in the supply of food, whether in the short term through food aid, or over time through production-enhancing strategies such as the Green Revolution. An FED mechanism, on the other hand, suggests that entitlements should be strengthened. In the long term this may involve fundamental shifts in the structures of access to endowments and livelihoods, while in the short term aid would be necessary either by income transfers, dumping food in quantities sufficient to reduce prices, the creation of food-earning opportunities, or promoting a free press to publicize events and prevent inaction. In practice, either just food or food *and* cash have formed the basis of immediate famine relief. A case has been made for the advantages of money rather than food in the hands of the vulnerable.[13]

The 1980s and early 1990s have seen a growing frustration with the inability of international humanitarian relief to gain access to civilian populations caught in civil wars or isolated in some other way. UNICEF and other UN bodies worked hard in the case of Sudan to negotiate 'corridors of tranquillity' through which they could transport relief food to populations in the south of that country (Minear 1991). Similarly, the Kurdish minority in Iraq remains isolated to a large degree. This has led to discussion in a number of countries of when and how national sovereignty should yield to humanitarian care for people in disastrous situations. This problem is more than a relief policy issue, and raises geopolitical and, above all, ethical concerns about the human right to food, shelter, and health care (de Waal 1991; Eide *et al.* 1984).

Early warning systems

One of the tools that has been developed since the Indian Famine Code and particularly since the famines in the Sahel in the 1970s has been famine early warning systems. Hervio (1987) defines them as 'instruments for measuring variations in the factors that determine a population group's food situation, measuring the consequences of these variations and finding specific solutions which will prevent famine occurring'. There are many different approaches to such systems, and in the view of some far too many. For example, there are about forty in the whole of Africa, with at least six rival systems in Mali alone. There is the FAO's Global Information and Early Warning System developed in the 1970s which predicted crop yields by estimating biomass from satellite imagery. This system suffered from the same intellectual ancestry as its food security programmes. Others work on the principle of a food balance sheet, where food production minus food needs equals imports plus food aid. The problems of pitching the scale of analysis at the national level and not incorporating the essential explanatory element of *access* have limited the usefulness of this type of early warning system. More sophisticated systems have since been developed, although the practicality of collecting the range of data needed for an ideal system is doubtful (Nichols 1988). Various types of indicators of famine have also been used as additional warnings (Cutler 1985).

Strengthening livelihood systems

Moving to less direct measures for preventing famine, the strengthening of rural livelihoods is the most obvious policy requirement. Of course, these measures can have objectives other than famine prevention, and have development objectives that are justifiable as ends in themselves. A comparison of India and Africa with regard to macro-policy concerning rural areas has been made by Harriss (1988). She reaches the conclusion that the situation in India is so different that virtually no lessons can be transferred to Africa. In India, the government has attempted land-tenure reform, improved agricultural production technology (including irrigation), encouraged better processing and storage, operates a fairly effective public distribution system of food grains, and has a well-tried decentralized emergency response mechanism. These have meant that a reasonably effective famine prevention strategy has emerged. However, this list of achievements apparently has not led to the abatement of widespread malnutrition in India. Nonetheless, it becomes difficult to demonstrate the level of impact of different rural development policies in famine prevention, except in general terms. This is because the causes of famine are essentially conjunctural and always involve more immediate and local factors.

Response to famine from the grass roots

All the policy options discussed so far are 'top-down'. Complementary and sometimes conflicting responses to the trigger events that may bring famine can also be seen at the grass roots. In addition to household and community level coping, discussed in Chapter 3, various non-governmental organizations have attempted many of the same interventions just discussed, albeit 'from the bottom up'. Organizations close to the grass roots are well placed to provide certain kinds of information for early warning systems to complement the coarse data provided by satellite. For example, such systems have been set in place by organizations like Save the Children Fund and the Red

Crescent Society. These systems are run by local citizens, who compile information on food and livestock prices and sale of assets by households, among other indicators of food emergency (York 1985; Cutler 1984).

Likewise, many NGOs have found themselves providing famine relief, sometimes in parallel to large-scale governmental and international aid, sometimes as a conduit for some of such aid, sometimes alone in isolated areas that do not receive other forms of aid. Some NGOs have tried to administer food relief in ways that strengthen local livelihoods in the long run (M. Scott 1987). For instance, Operation Hunger in South Africa faced very large-scale hunger in the so-called Bantustans during the drought of 1991–2. They supported women's irrigated gardening groups as well as providing conventional mass feeding.

The federation of village development (*naam*) groups in Burkina Faso and other Sahelian countries known as the 'Six S' has sponsored grain storage on a co-operative basis. In this way farmers are released from the cycle of indebtedness caused by having to sell at a low price every year to a trader, followed by purchase of grain from that same trader at a higher price later during the 'hungry season' (Adamson 1982). The Gramin Bank in Bangladesh serves a similar function, providing credit for the 'uncredit-worthy' in order to start up income-generating activities. The conclusion we reached earlier that FED is generally more important than FAD as an explanation of famine suggests that extra income would be especially important in preventing famine, and that such schemes are capable of altering the pattern of access to enable people to improve long-term entitlements, and thereby reduce the risk of facing famine.

Numerous other efforts to strengthen livelihoods have been directed towards rural people vulnerable to drought. One of the most interesting of these was undertaken by the British charity War on Want in Mauritania following the Sahel drought and famine of 1967–73. Rather than spend all funds collected during the crisis on immediate relief, they launched a thorough study of the existing livelihood system of the people living in the Guidimaka region of Mauritania bordering the Senegal River (Bradley, Raynaut, and Torrealba 1977). On the basis of this study they proposed a long-term development programme centred on and controlled by village associations. The programme featured low-cost improvements of existing agricultural practices, which already included very sophisticated flood-retreat agriculture and careful livestock management. They attempted to diversify the livelihood system through the addition of fruit trees and craft industries. Their's was one of the first programmes to introduce small bunds (embankments) to increase infiltration of scarce rainfall into the soil – an innovation that later became very commonly promoted throughout the Sahel (Twose 1985; Harrison 1987).

CONCLUSION

The causes of famine are far too complex for single theories to claim universal applicability. The iterative access model allows rigorous theory to operate, but under conditions which each particular famine determines. Understanding each famine in this way makes it possible to evolve a more flexible analytical method for explaining famines, which in turn may assist more responsive and flexible policy.

NOTES

1 A large literature traces the role of hunger in history. M.M. Cohen (1977) overviews the evidence of hunger in prehistory. The recorded history of hunger is traced by De Castro (1977); Sorokin (1975); and Arnold (1988); for the Graeco-Roman world (Garnsey 1988) and ancient China (Mallory 1926). There is a recent thorough interdisciplinary view edited by L.F. Newman (1990).

2 Such policy-induced famines include those associated with collectivization in the USSR in the 1930s and the introduction of the Peoples' Communes system in China in 1958. There may also be 'policy famines' as a result of inappropriate state interventions in the event of disruption of food systems, as with the Irish famine after 1840, when the British government adhered to a *laissez-faire* policy of limited intervention, with restricted food aid (imported American maize) that could not be eaten locally (Woodham-Smith 1962).

3 Meillasoux (1973, 1974) and Copans (1975, 1983) are typical of the analysis at the time of famine in Sahelian West Africa. Franke and Chasin (1980) reviewed and summarized much of the French literature of the period. Elsewhere in Africa Wisner and Mbithi (1974) and Wisner (1980) had much the same thing to say about famine in East Africa, while Bondestam (1974) and Kloos (1982) elaborated a 'political economy' explanation for an area in Ethiopia (the Awash Valley), and Hussein (1976) extended the analysis to Ethiopia as a whole. Similar analysis of the role of class relations in undermining traditional drought-coping methods and in causing environmental destruction (which made the land more prone to the effect of drought) also appeared for southern Africa (Cliffe and Moorsom 1979; cf. Wilmsen 1989).

4 Those who introduce social relations into their work on land degradation include Blaikie (1985b); Blaikie and Brookfield (1987); Little and Horowitz (1987); Hecht and Cockburn (1989); and Juma (1989).

5 War and famine have been treated by a thematic issue of *Review of African Economy* No.33, August 1985. Dando (1981) provides a broad historical overview of cases where war was a cause of famine. Shindo (1985) focuses on the indirect influence of government military expenditure in creating famine conditions. Brandt (1986) and E. Hansen (1987) also take up this theme. Jacobs (1987) discusses the blockade of Biafra (southeastern Nigeria) that produced a million famine deaths in the late 1960s. Lemma (1985) treats war and famine in Ethiopia; Minear (1991) deals with southern Sudan; S. Smith (1990) summarizes the effect of war in Mozambique and Angola. Focusing on other parts of the world, Kiljunen (1984) writes about Kampuchea during its 'decade of genocide'; the so-called Great Bengal Famine, caused in part by war-time economic policy, is discussed by Aykroyd (1974) and Sen (1981).

6 Additional depth on northern Nigeria is provided by Mortimore (1989). Similar treatments of famine's 'chain of causes' is provided for Malawi by Vaughan

(1987), for Kenya in 1971 by Wisner (1978b, 1980), for Kenya in 1984 by Downing, Gitu, and Kamau (1989), for Ethiopia by Mariam (1986) and Kebbede (1992).

7 De Castro contributed more than half a century of famine studies that deviated from the conventional wisdom by emphasizing social relations. A Brazilian medical doctor and geographer, De Castro began documenting chronic under-nutrition and what we would call today 'vulnerability' in Latin America in the 1920s. In the early days of the United Nations he served in the FAO nutrition division, but became disenchanted by international efforts to prevent famine. There is a bibliography of De Castro's major works on famine and a critical review in Wisner (1982); the best-known works include *Geopolitics of Hunger* (1977) and *Death in the Northeast* (1966).

8 Summaries of these breakthroughs in natural and physical science can be found in a recent collection edited by Wilhite and Easterling (1987).

9 Much has been learned by field administrators and government officials over the last two decades. There is a body of experience on administration of famine refugee camps (Harrell-Bond 1986), provision of food aid (Drèze and Sen 1989), and early warning systems (Walker 1989). See also Borton (1988) for an excellent review of recent British experience with famine and emergency relief. Jackson (1982) and Crow (1990) are more critical of the aid process, especially the provision of food aid. R.C. Kent (1987a) has summarized a large body of increasingly critical commentary on relief aid in many forms.

10 'The underlying cause of famine is crop failure which undermines incomes of the already poor' (Mellor and Gavian 1987: 539). Platteau (1988) emphasizes the importance of declining aggregate food production in Africa, but attributes it more to backward agricultural technology and to pricing policies, as do many other authors. For the alternative emphasis, especially on the influence of govern-ment policy (e.g. pricing, credit, marketing, infrastructure, research and develop-ment) on African food production, see Bates (1981); Berry (1984); Raikes (1988); Wisner (1988b); Odhiambo *et al.* (1988); Cheru (1989); Bernstein (1990); Achebe *et al.* (1990); Rau (1991); Wisner (1992b).

11 Drèze (1988: 19) says that exports abroad from India were frequent during famines in the nineteenth century.

12 Southern African Development Coordinating Conference, created in 1980 by Mozambique, Angola, Zimbabwe, Botswana, Swaziland, Zambia, Malawi, Lesotho, Tanzania (and now including Namibia). On their common food policy see Prah (1988); Rau (1991: 125–8); Wisner (1992b).

13 On making food aid more effective, see Singer, Wood, and Jennings (1987); M. Scott (1987); Cohen and Lewis (1987); Hopkins (1987); and Raikes (1988).

5

BIOLOGICAL HAZARDS

INTRODUCTION

So far we have introduced the notion of 'vulnerability' and shown its relation to a set of processes that affect what we called 'access'. In Chapter 4 we explained that famine is not synonymous with drought or other natural conditions, nor with a decline in food availability as such. What counts is people's vulnerability to disruptions in their access to resources and livelihoods.

The present chapter continues that analysis in regard to 'biological disasters' that affect both people (disease) and their crops or animals (disease and pest infestations). It examines a number of events from early civilization, from the middle ages, and recent times to see if here, too, we find differences in vulnerability at work. Several connections will be made with health problems that are not specifically disasters, since health is a crucial aspect of vulnerability in general. It seems most convenient to incorporate them here in the wider discussion of biological factors.

The discussion includes the Plague of Justinian, an outbreak of bubonic plague contemporaneous with the fall of the Roman Empire, and the pandemic called the Black Death that wiped out about a third of the European population in the fourteenth century. Contemporary crises examined include the AIDS pandemic and the less dramatic biological challenges produced by the resurgence of many vector-borne diseases, sanitation problems in Third World cities, and depletion of wild genetic reserves.

Biological hazards include micro-organisms such as those responsible for epidemic human disease, epizootics such as rinderpest and swine fever, and plant diseases. Insects and other animals can transmit disease (mosquitoes, rats, lice, fleas) or can destroy crops (dioch birds, locusts, army worms, grasshoppers). As noted in the case of potatoes in the nineteenth-century famine in Ireland, normal biological populations are usually genetically diverse, so that biological hazards rarely trigger disasters.

To varying degrees, human systems have also developed social (as well as biological) resistance to such risks. Human culture has also developed ways of tolerating crop and animal losses up to certain levels. Indeed, people may be ambivalent towards certain pests (the grasshoppers in southeastern Nigeria that constitute the major crop hazard are seen as a food source by women and children (Richards 1985)). Elaborate adjustments of techno-social systems have developed in the face of diseases of plants and animals, and of crop losses due to pests and vermin (Mascarenhas 1971) and post-harvest losses (Bates 1986). In most cases the existence of 'normal surplus' is enough simply to absorb such losses. That is, during 'normal' times the tendency of subsistence systems is to *overproduce* beyond subsistence needs as a built-in, structural insurance that needs will still be met in all but the very worst times (Allan 1965; Porter 1979). What has been common practice by peasant farmers and pastoralists for centuries has recently been redis-covered in the context of European and North American agriculture under the rubric of 'integrated pest management'. A degree of loss is tolerated until it exceeds the marginal cost of action against the pest (Altieri 1987).

Disaster literature, with the exception of specialized public health writing, has tended to neglect biological hazards.[1] Early disaster research was obvi-ously confused when facing the richness of the biosphere. Burton and Kates (1964) included everything they could think of in their category of 'biologi-cal hazard', including athelete's foot (a minor fungal problem). We propose to take a more systematic look at people and their livelihoods in relation to biological hazards. Biological disasters are analysed in their own right (whether as risks to human health or diseases of crops and animals). But in addition, an understanding of biological hazards will permit better compre-hension of the health problems integral to vulnerability to many other types of hazard. It also allows understanding of the significance of health in the impact of other hazards, where, for example, floods expose people to new health risks.

Limits to vulnerability analysis?

In Chapter 1 we cautioned that differential vulnerability is not present in all disasters. Although it is relevant in most, there are bound to be 'limiting cases', where vulnerability resulting from social structures is of little signifi-cance to the type or intensity of hazard. In such cases, the issue of how a hazard's impact is distributed among this or that human subgroup would be irrelevant. The Black Death and AIDS may well be such cases, pandemics that are perhaps no respecters of the social class or other differential charac-teristics of the population. This chapter will explore such limits to the applicability of 'vulnerability' as a concept in relation to biological disasters. Severe and widespread plagues, pestilence, and infestations may serve as an extreme test of the vulnerability concept.

But aside from this concern, there are other boundaries involved in biological hazards. The significance of human health problems as part of other disasters is apparent in other chapters from our analysis of vulnerability to hazards such as floods, drought, and tropical cyclones. These have potentially severe biological and epidemiological consequences. For instance, flooding is commonly followed by epidemics of diarrhoea due to contamination of drinking-water sources. There is no doubt that people who survive the immediate impact of some hazards can suffer much more from a consequent health crisis. Conversely, vulnerability to the impacts of many hazards is increased in populations with chronically bad health. Hence an understanding of disease helps us grasp the implications of access (to adequate resources and livelihoods) in understanding the full impact of flood, drought, and other natural events.

Earthquakes, landslides, and tornados are said to carry no acute or secondary epidemiological threats themselves. Occasionally the disease consequences of some hazards such as earthquake have been overestimated (Cuny 1983: 44–9). Others (floods, cyclones, tsumanis) can lead to increased water- and vector-borne diseases such as typhoid, cholera, leptospirosis and typhus, malaria, and encephalitis. In such situations, environmental engineering and surveillance are generally advised rather than mass immunization (PAHO 1982: 13–21, 53–60). It is important to note that the health consequences of a hazard itself may be insignificant compared with those of removal to disaster relief camps. The high population densities in refugee settlements are often responsible for the increased spread of communicable disease, and sanitation is often rudimentary (PAHO 1982: 3–12; Simmonds, Vaughan, and Gunn 1983: 125–65).

THE PERSPECTIVE OF POPULATION BIOLOGY

Microbes, disease vectors, and human beings have evolved together over hundreds of thousands of years. Our immune system derives from that developed through the interaction of our primate ancestors with the environment over millions of years (McKeown 1988; Anderson and May 1982). Given the time involved, human 'adaptation' to biological challenges ought to be more thorough than our purely cultural adaptations to drought or architectural safeguards against earthquakes. Even if humans had not evolved a generalized immune response, genetic polymorphism would provide sufficient diversity in populations to minimize the risk of catastrophic mortality (Ruffié 1987).

Yet human history is punctuated by disease disasters. Marks and Beatty (1976: 3–18) find written evidence of epidemics from at least the fifteenth century BC, with the most severe involving tens of millions of deaths. The common factor in many of these epidemics of increased human mobility is striking. The Plague of Justinian (AD 541–93) took place during the decline of the Roman Empire, accompanied by numerous wars and population movements. The Black Death in Europe (1348–1400) corresponds to the increase of trading contacts said to have preceded the birth of capitalism. The most recent, a great influenza pandemic (1918–20), accompanied the extreme displacement of people during and after the First World War.

The role of human migration is also highlighted by the conquest of the 'New' World. New disease organisms such as those responsible for smallpox and measles were introduced into the Caribbean and Latin America from the sixteenth century. They led to millions of deaths among the indigenous populations in a matter of years (Crosby 1986: 195–216; Marks and Beatty 1976: 160–4). European conquest of Australia and New Zealand, previously isolated in a similar way from Old World pathogens, was accompanied by demographic collapse of the indigenous populations (Crosby 1986: 217–68; 309–11).

Why does catastrophic mortality on this scale persist in large, genetically diverse human populations with highly developed immune systems? Why at times have millions of people died in a matter of a few years? The answer lies in the action of many variables leading to vulnerability to the diseases in question. Epidemiology has used the concept of 'risk' for many years (MacMahon and Pugh 1970). The classic model of epidemiological causation involves the interaction of disease agent, the environment, and the host. The host is often found to be differentially 'at risk' because of genetic make-up, age, gender, and social class, and the influence some of these may have on the particular environments that are lived or worked in.[2] Also, chronic under-nutrition and disease are known to interact in a mutually reinforcing manner (Scrimshaw, Gordon, and Taylor 1968). In this way, the 'host's' nutritional status cannot be separated from the status of sanitation in the 'environment' (Cairncross 1988; Wisner 1988b: 87–111).

This chapter explores whether differential access to the means of attaining and maintaining 'health' is at work in the most extreme health disasters as well as in ordinary day-to-day life. In general there are three lines of defence against such hazards: genetic, environmental, and cultural. When one or more of these are compromised, usually by the slow build up of multiple stresses (or root causes and dynamic pressures as described in Chapter 2 in the PAR model), disaster can result.

LIVELIHOODS, RESOURCES, AND DISEASE

To begin with, we explore the structure of a 'biology' version of the access model developed in Chapter 3. Initially this will relate to the way in which normal 'access to health' is affected by the social, economic, and political factors that determine access to resources and livelihoods. This is then extended to examples of biological disasters in order to discover how successful the notion of vulnerability is in analysing catastrophic events.

The role of access

We can use the basic access model presented in Chapter 3 (Figure 3.1) to think about health issues. Boxes 1, 2b, and 4 determine household and individual resources and income opportunities. Basically, the household has access to the major objects of productive land, labour, tools, and livestock with which to provide a livelihood with 'health' (nutritional level, housing and sanitation, health care, etc.) at a certain status. This level is attained by using resources in an array of income-earning activities (box 3a) and then disposing of household income (box 7). Some of these activities are hazardous in other aspects of health. Pesticides may be applied without people being adequately protected by gloves, masks, boots, etc. Many 'development' projects, among them irrigation schemes, have associated health risks, including malaria and schistosomiasis (bilharzia) (Hughes and Hunter 1970; Wisner 1976a; Bradley 1977). Wage-migration may put a family member in contact with disease vectors (see Forde (1972) on contact with tsetse flies, Prothero (1965) on anopheline mosquitoes, etc.) or sources of infection not found at home.[3] In this way, access to resources that are meant to sustain family health is possibly achieved only at the cost of exposure to health-degrading circumstances.

'Normal' levels of morbidity and mortality are in part determined by the outcome of these processes. The household's reproductive fertility is decided through a complex chain of cultural, psychological, and physiological events (boxes 8 and 9), which in turn influence the health of individual members of the household through the sharing of food, the antenatal health of children, and the well-being of mothers after childbirth.

'Normality' (often not 'healthy', but relatively static) can contribute to biological disaster in three ways. Firstly, there may be sudden changes in class relations or structures of domination (boxes 1 and 4) that reduce access to resources essential for the maintenance of even minimum 'normal' levels of health. Expropriation of land is an example of such change. Or there may be a breakdown in law and order that denies resources, increases uncertainty of future income, or propels people into headlong flight. When displaced people come into contact with disease vectors or agents new to them, widespread death can occur (Hansen and Oliver-Smith 1982). In these examples, biological disaster occurs as a sequel to social disruptions such as war, pogroms, dispossession, and famine.

Another explanation of biological disasters involves the occurrence of epidemic (or epizootic) disease that provokes further breakdowns in health. In such cases the 'vicious cycle' of vulnerability is very apparent. For instance, in the late 1970s an epizootic of foot-and-mouth disease among cattle of the pastoral Maasai in Kenya led the government to ban the sale of these animals (because of fear that Kenyan beef would be banned from European markets). The quarantine blocked a potential income-generating activity. This undermined Maasai livelihoods, with consequent increased human morbidity and mortality. Their increased vulnerability thus arose from a sudden decline in nutrition, further complicated by tuberculosis and measles, derived from the economic shock of government policy (D. Campbell 1987).

Epidemic outbreaks of human disease can have the further effect of diverting household income through increased medical, ceremonial, and funeral expenditure. A weakened or demographically-depleted household labour force will command less income. Barnett and Blaikie (1992) report that AIDS in Rakai District, Uganda has involved people in so many funerals (where work must be suspended for at least three days) that they must work by moonlight and in secret so as to be able to complete essential agricultural operations in time. Death or weakness of livestock essential for traction in farming can also contribute to the deteriorating ability to maintain human health following epizootics.

VULNERABILITY-CREATING PROCESSES

Since we want vulnerability analysis to be useful to planners and other development workers, we need to specify as carefully as possible how chains of social and economic events 'translate' various factors so as to generate people's vulnerability to biological hazards. We do this by looking at various levels of social and natural environments.

The micro-environment

Diet, shelter, sanitation, and water supply work together at the household level to determine vulnerability to biological disaster. The synergism linking disease resistance and nutrition has already been mentioned. During the drought in the Sahel (1967–73), most of the 100,000 lives lost were due to the interaction of starvation and measles (Morris and Sheets 1974). Baseline nutrition as a factor in famine has already been discussed in Chapter 4. Previously well-nourished populations, such as the Dutch in the latter months of the Second World War (1944–5), are able to survive famine conditions that would surely kill others in far greater numbers.[4]

Vulnerability may be specifically affected by the type and location of housing. The location, especially in urban areas, is constrained by law, land prices, distance to livelihood, and availability of building materials. The million or so living inhabitants of Cairo's 'City of the Dead' (originally cemeteries), or those who inhabit the garbage dumps of Manila and Mexico City did not choose these locations on account of the healthy ambience. The resources and income opportunities that dictate housing also determine the quality of water and sanitation. In later chapters we will show how similarly constrained locational decisions increase vulnerability to mudslides, floods,

106

storms, and earthquakes. Biological hazards clearly become enhanced by various social and economic factors, especially those affecting house type and location.

Regional social environment

The impact of migration on people also plays an important role in determining vulnerability. Roundy (1983) has shown that movements of Ethiopians even over small distances can introduce people to health threats when a large change in altitude or other major ecosystem boundary is involved. Seasonal wage-migration of highland dwellers to the coasts of central and Andean South America produces similar effects. Long-distance migrants can become victims of malaria and sleeping sickness among other diseases.

When whole families migrate to very different environments (willingly or with various degrees of state coercion) severe health problems have been recorded. Migrants from the northeast and south of Brazil suffer high morbidity and mortality in newly-settled Amazonian habitats. Migrants from the northern parts of Ethiopia (usually under powerful state coercion) to the southwest of that country also suffered a great deal (Kebbede 1992; Clay, Steingraber, and Niggli 1988). Similarly in the 1920s Soviet settlers in Siberia suffered epidemics (Pavlovsky n.d.). Longitudinal studies have revealed that several decades after being resettled from the valley flooded by the Kariba dam on the Zambia/Zimbabwe border, settlers were still suffering measurable health consequences (T. Scudder 1980; cf. Hansen and Oliver-Smith 1982). Forced displacement has had even more severe health consequences than migration.[5]

All such displacements have resulted in significant levels of epidemic disease. Even the best-run refugee camps have problems with sanitation (Harrell-Bond 1986: 244–8; Howard and Lloyd 1979; Khan and Shahidullah 1982; PAHO 1982). Outbreaks of epidemic diarrhoeal disease have been reported in Latin American camps (Isaza *et al.* 1980), and in Africa (Rivers *et al.* 1974) as well as Asia (Anton *et al.* 1981; Temcharoen *et al.* 1979). Cholera, though less frequent, has also been recorded (Morris *et al.* 1982). Crowding produces increased potential for airborne infections and other diseases transmitted by personal contact. Until it was controlled, smallpox could spread rapidly through non-immunized refugee populations (Mazumder and Chakrabarty 1973; McNeil 1979). Higher densities of population can also make the transmission of malaria easier, especially where there is drug-resistance (Reacher *et al.* 1980) or acute nutritional stress (Murray *et al.* 1978).

Regional physical environment

Degraded environments can reduce people's access to their livelihood resources, and so increase their vulnerability to biological and other hazards.

For this reason, soil erosion, desertification, and alkalinization have been labelled 'pervasive' or 'slow-onset' hazards (Pryor 1982; Blaikie 1985b). But environmental degradation can also affect vulnerability in other ways, by reducing or damaging the earth's genetic materials. Air- or waterborne industrial pollution can also have this effect. As well as rendering fisheries and farmland barren, it can cause the death of trees, wildlife, and mangroves (Eckholm 1976; Maltby 1986; Weir 1987), thus reducing livelihood resources, and also damaging biodiversity.

There is much discussion of worldwide threats to biodiversity (Wilson 1988; Juma 1989; Fowler and Mooney 1990). In the long run declining biodiversity could provoke or at least amplify biological disasters, especially as the wild ancestors of our food crops are lost or abandoned. Since roughly one-quarter of all pharmaceuticals have plant or animal origins, diminished biodiversity could undermine the scope of therapy in the face of new diseases. Depletion of regional genetic resources has already contributed (as 'trigger') to disasters, and should be considered a pervasive hazard, as the Irish famine shows (Box 5.1).

PRESSURES AFFECTING DEFENCES AGAINST BIOLOGICAL HAZARDS

To understand how occasional biological disasters occur, it is clear that we need to analyse many factors that may increase people's vulnerability to the impact of diseases of humans, plants, and animals, or to infestations of pests. In addition, we need to evaluate environmentally-damaging processes that diminish the availability of genetic material, thereby undermining food sources and medicinal repositories. We can consider these in terms of genetic, environmental, and cultural defences.

Genetic defences

Genetic polymorphism confers a degree of resilience in a population of plants, animals, or humans. These benefits can be defeated in a number of ways. Firstly, a new and virulent organism can be introduced unknown to an existing biological community. An example is the great loss of life due to smallpox when it was introduced into the New World.[6] Similarly, hogs, horses, cattle, and the black rat thrived once they were imported to the New World, and pushed competitors out of ecological niches (Crosby 1991). The cattle herds (*Bos indicus*) of eastern and southern Africa in the 1890s were reduced by up to 90 per cent by rinderpest, new to this population. Their hapless owners, weakened by the economic stress of losing so many animals, and suffering social disruption due to German and British colonial expansion, also confronted new disease organisms brought from West Africa: guinea worm and jiggers. Resulting mortality was probably one of the major

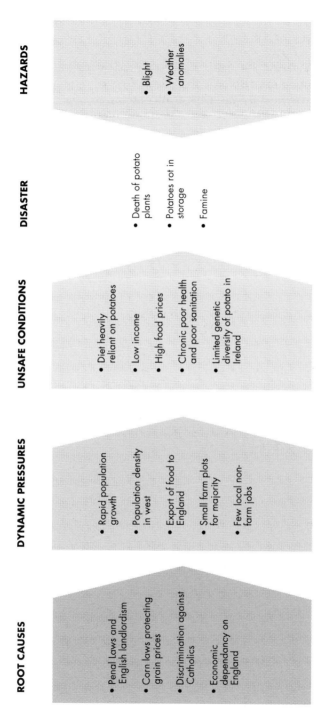

ROOT CAUSES

- Penal Laws and English landlordism
- Corn laws protecting grain prices
- Discrimination against Catholics
- Economic dependancy on England

DYNAMIC PRESSURES

- Rapid population growth
- Population density in west
- Export of food to England
- Small farm plots for majority
- Few local non-farm jobs

UNSAFE CONDITIONS

- Diet heavily reliant on potatoes
- Low income
- High food prices
- Chronic poor health and poor sanitation
- Limited genetic diversity of potato in Ireland

DISASTER

- Death of potato plants
- Potatoes rot in storage
- Famine

HAZARDS

- Blight
- Weather anomalies

Figure 5.1 'Pressures' that result in disasters: the Irish Potato Famine 1847

Box 5.1   The Irish Potato Famine (1845–8)

This tragedy had many contributing causes (Woodham-Smith 1962; Walford 1879; O'Brien and O'Brien 1972; Aykroyd 1974; Sen 1981; Regan 1983). But reliance by the impoverished rural workers for 90 per cent of their food energy on potatoes with very little genetic variation meant that both the potatoes themselves and the people were highly vulnerable to the blight caused by the organism *Phytophora infestans*. The root causes, pressures, and vulnerability effects at work in this biological disaster are illustrated in Figure 5.1, a version of the PAR model.

The root causes of this famine go back to the 1650s, when Cromwell's conquest of Ireland began systematic discrimination against the Catholic peasant majority. Firstly, a large number were forced to resettle in the west, creating a population distribution that two centuries later gave high rural densities and small farms. This encouraged reliance on potatoes because they could yield up to 6 tonnes per hectare (2.5 tons per acre). The Penal Laws (1695) reinforced this pattern as they made it mostly illegal for Catholics to own land. English absentee landlords owned most of the land in much of Ireland. Protected by the Corn Laws, landlords were able to export grain and meat to England profitably without competition from foreign grain that would have forced the price down.

In the many non-blighted years the potato had provided increased dietary energy leading to higher child survival rates. This enabled an increase in the poor Irish Catholic population. There was pressure to produce large families both to work the land and, above all, to seek paid employment (a part of which money would be remitted to the rural home) in British cities and in America. Population growth exacerbated pressure on land in the west and maintained the trend towards very small farms. Export of grain and other food to England further reinforced reliance by the majority on potatoes. Ireland's economic dependency on England produced an additional pressure on subsistence potato farms because the resulting low rate of domestic (Irish) saving meant that few non-farm jobs were available locally as an alternative.

---

reasons for the collapse of militant resistance to colonialism in the region (Kjekshus 1977).

Even when there is some resistance to a disease agent in a population, sufficient time may have elapsed since the last outbreak to produce a large number of non-immune individuals. European history is thus punctuated with catastrophic epidemics of bubonic plague (transmitted by fleas living on rats) even though it was endemic all along.[7] In addition to a non-immune population of sufficient size, pre-existing spatial networks to channel the progress of infection and a high level of human movement along these networks are required.

Thus, the Plague of Justinian apparently originated in Egypt and spread along trade routes. These were swollen with soldiers and refugees of the wars then being fought to regain parts of the Roman Empire lost to the

Box 5.1 continued

Specific vulnerability effects included a diet heavily dependent on potatoes, with reports that an adult would consume up to 6 kilograms (13 pounds) a day (Aykroyd 1974: 32). At the time the 'Great Starvation' became fully manifest, one-half of the Irish population was dependent on the potato (Regan 1983: 114). Besides a hog, fattened on kitchen scraps for sale to meet taxes and other monetary expenses, potatoes and peat to burn for heat were the mainstays of the peasant economy.

Vulnerability was further enhanced by low cash incomes, making the purchase of alternative foods very difficult. Because of the Corn Laws, prices of alternative foods were always high, and, of course, very much higher during the famine (recalling FED, discussed in Chapter 4). Crowding, poor housing, and poor sanitation added to vulnerability by undermining health status, providing less resistance to the physiological effects of hunger and cold.

Finally, the potatoes used by the Irish peasantry were all descended from a few hundred tubers brought from the Andes in the early 1600s. By the 1840s the potato gene pool in Ireland was extremely homogeneous, thus vulnerable to widespread damage by disease. Genetic diversity would have conferred at least some protection, by increasing the chance that the gene pool contained some potato varieties resistant to blight.

Looking again at Figure 5.1, note that the physical hazard side must include not only the disease agent *Phytophora infestans*, but also the exceptionally warm weather in 1845 that triggered the explosion of blight, and the unusually cold winter weather of 1846–7. The cold is said to have killed many who had been weakened by hunger during the previous year. (Such sequential impacts of different hazards – drought followed by flood, tornado followed by flood, earthquake followed by freezing weather – are often the triggers of serious disasters.)

The effect of this biological disaster was at least 1.5 million dead between 1845 and 1848, and another 1.5 million forced to emigrate. The 1881 census revealed an Irish population still 3 million people less than the pre-famine total.

Vandals and Ostrogoths. Bubonic plague appeared in Byzantium, capital of the Eastern Empire, in AD 542, and by the end of the century this city had lost half its population. Following a second route along the Mediterranean, it appeared in France in AD 543. Depopulation was so great that much land fell idle, taxes were not paid, and estates were widely replaced in the seventh century by a pattern of small freeholders (Russell 1968).

Catastrophic economic loss of crops to disease or pests can be avoided for a time by heavy use of protective agrochemicals. But there are serious problems involved in this strategy, not least the foreign exchange crisis faced by many countries forced to import chemicals and export cash crops. Pests and diseases often develop resistance to the chemical agent, and competing or beneficial organisms may be killed. This produces the well-known pesticide 'treadmill' of resistance and resurgence of the pest (Debach 1974; Altieri

1987). For example, cotton on the vast irrigated Gezira scheme in Sudan began to be cultivated in the 1920s. In the early 1950s the crop was sprayed with pesticides once or twice. By the early 1980s cotton was sprayed nineteen times during the growing season in order to control a greater number of pests. Widespread use of agrochemicals, especially DDT, to combat malarial mosquitoes has produced pesticide-resistant mosquitoes, with the result that malaria is resurgent in many parts of the world (Chapin and Wasserstrom 1981; Sharma and Mehrotra 1986; Learmonth 1988: 208–11; Matthiessen 1992).

Expansion of export crops into forest, fallow land, and plots previously devoted to subsistence crops has caused the extinction of many local varieties of legumes and other food crops as well as many gatherable forest products (Juma 1989). For example, the urban market for red kidney beans in Kenya has caused farmers to stop using many indigenous varieties of legume and concentrate on *Phaseolus vulgaris* because they need cash. These genetic resources are lost to future generations who might use them to strengthen livelihoods and to reduce disaster vulnerability.

Environmental defences

The single most important environmental defence against catastrophic disease and infestation is probably dispersed settlement. Rapid urbanization, whether as the result of rural insecurity (as during the decline of Rome and early middle ages) or due to the rise of capitalism (from the fifteenth century), is also associated with increased vulnerability to epidemics.

Potable water and sanitation systems are also a factor. There was probably little significant difference in sanitation between rich and poor in Europe until the nineteenth century.[8] Today in Third World cities (many with double digit rates of urbanization) the haves enjoy indoor plumbing while the have-nots are lucky if they have a communal water standpipe within a few hundred metres of their front door (Agarwal *et al.* 1989; Feachem, McGarry, and Mara 1978; Cairncross, Hardoy, and Satterthwaite 1990a). Many mega-cities (e.g. Calcutta, Lagos, Mexico City) have sanitation systems based on drains and water mains at least 100 years old. Many, like Howrah (a city of 2 million across the Hooghly River from Calcutta) have no sewers at all.

Different classes of people were also more or less able to get access to safe havens and flee from the Black Death. While it was not guaranteed protection, escape to country houses from affected cities was an option only the rich enjoyed (Ziegler 1970). The spatial organization of residence in the mega-cities of Africa, Asia, and Latin America is likewise significant in producing differential vulnerability to disease. The dangers are increased with declining budgets for maintenance of even minimal urban infrastructure in many countries. Reduced public expenditure is often due to

IMF/World Bank insistence on austerity programmes in the face of foreign debt (Hardoy and Satterthwaite 1989; Cairncross, Hardoy, and Satterthwaite 1990a). In Chapter 2 we introduced this as one of the major 'root causes' of disasters.

Cultural defences

We have discussed the diverse ways people tolerate crop and livestock losses in the coping sections of Chapter 3. One of the most important of these is to combine, if possible, an assortment of livelihood activities (see the section on coping in Chapter 3). Poor farmers not only try to grow a variety of crops, usually intercropped, but they may engage in a variety of non-farm activities, including trade, craft production, and services (Chambers 1983; Guyer 1981; Wisner 1988b).[9] Current 'development' efforts often incorporate rural households into commodity production in a manner that reduces the diversity of rural livelihood opportunities (Bernstein 1977, 1990; Wisner 1988b: 187–97). Failure of a crop under such circumstances can have catastrophic results.

ROOT CAUSES AND PRESSURES: BIOLOGICAL HAZARDS AND VULNERABILITY IN AFRICA

In our PAR model, 'root causes' of vulnerability are found in global economic processes events, and 'dynamic pressures' are to be found in the structure of particular societies. This section examines some of the linkages between people's vulnerability to biological hazards and the political and economic root causes and pressures that explain it. The illustrations are drawn from Africa, and a case-study of AIDS is given in Box 5.2. Since the era of independence of the 1960s, which prompted such high expectations for the people of most Sub-Saharan African countries, the Four Horsemen of the Apocalypse have ridden over nearly every part of the continent. War and famine have affected nearly two dozen of these countries and have interacted in numerous ways with the other two biblical threats, disease and pestilence.[10] During this period up to 30 million people have been displaced. At the peak of the most recent acute crisis (1983–5), the official total of refugees was 5 million (CIMADE, INODEP, and MINK 1986). The resulting disruption of livelihoods, and accommodation in makeshift camps, have contributed to high levels of civilian deaths. Meanwhile, drought, flood, and cyclone affected an already weakened people, and locusts, Sudan dioch birds, livestock diseases, cholera, and measles took their toll (Harrison 1987; Stock 1976; Timberlake 1985).

The broader context of these tragedies is a failure of social and economic development. Many African nations lost ground during the 1980s, in comparison with their position in the 1960s or 1970s, in respect of infrastructural development, services, and productivity per capita (Cornia, Jolly, and Stewart 1987; Whitaker 1988; Rau 1991). Foreign indebtedness is high when compared with the ability to service debt, and the dependency on foreign aid is as much as it has ever been (George 1988; Onimode 1989).

African governments have cut their health budgets because of loss of income from exports (whose prices on world markets consistently fell during the 1970s and 1980s) and because of financial austerity programmes mandated by the IMF and World Bank (Wisner 1992a). Maintenance of infrastructure, procurement of medicines, training, and plans to improve primary health care have all suffered. Lack of road maintenance and shortage of foreign exchange to import fuel and spare parts for vehicles have meant that mobile services to isolated villages have been interrupted. Such services had been very effective in providing early warning of famine and epidemic disease hazards (see Chapter 4).

Nigeria provides a useful illustration of the links back from unsafe conditions, through pressures to root causes. The petroleum boom of the 1970s weakened agricultural and other rural livelihoods in Nigeria (Watts 1986). Economic crisis in the 1980s eroded the purchasing power of cash wages, and many people lost employment. The onslaught on livelihoods was structured by long-standing gender, tribal, and class biases. Rural–urban differences in access to health care and income differences between classes had always been great. The IMF structural adjustment programme increased them (Nafziger 1988: 123–4; Wisner 1992a: 152). Women have very heavy work burdens and very poor health. In the north maternal mortality (an index of both health care and 'normal' health) is 1,500 per 100,000, compared with 150 in Zimbabwe and five in Europe (Wisner 1992a: 161).

In 1991, half of the worldwide deaths from cholera (7,200) occurred in Nigeria. The deteriorating health care system combined with increasing individual vulnerability required only the presence of the cholera organism (the bio-physical 'hazard') for a disaster to occur. The spread of cholera reflected the problems of spatial structures (infrastructure, patterns of urbanization) and spatial processes (frequency of public markets, festivals, etc.) (cf. Stock 1976).

STEPS TOWARDS RISK REDUCTION

Earlier successes

Successful campaigns against certain illnesses have been recorded which have operated both from 'the top down' as well as 'from the bottom up'. Control of yaws just after the Second World War and the more recent eradication of smallpox were successes in large-scale biomedical administration. The disease agents and mode of transmission in both cases were straightforward. Treatment for yaws with penicillin and vaccination for smallpox were

114

uncomplicated, and follow-up was not necessary. No insect vectors with complicated life cycles were involved. UNICEF is currently in the midst of a more ambitious worldwide effort to immunize all children against tuberculosis, tetanus, measles, diphtheria, pertussis, and polio. Some of the vaccines (measles, for instance) must be kept cool until injected, and so require a complicated 'cold chain' of refrigerated facilities in storage and transport. Others require follow-up immunization (polio, for instance). There is some debate as to whether malnourished children with depressed immune systems can form antibodies in response to immunization in some cases. (This did not affect the anti-smallpox campaign because the immune reaction against live cowpox antigen is universally strong.) Another partial 'top-down' success against a biological hazard involves locusts in Africa and the Middle East. Aerial surveillance (conditional on international agreements with regard to air space) and massive aerial spraying of locust breeding sites with insecticide seem to have been moderately successful.[11]

On the other hand, the biggest increase in human longevity came as the result of improved food supply, housing, and sanitation in nineteenth-century Europe and North America, not through biomedical intervention at all (McKeown 1988). Security against biological hazards requires some minimal level of access to resources (as described in the access model above).

Furthermore, a number of the potential health disasters facing us are not suited to the 'top-down' approach. AIDS, cholera, plague, malaria, as well as other complex biological threats such as deforestation, desertification, and the loss of species cannot be tackled by 'top-down' problem-solving. Instead detailed local knowledge of highly variable social and environmental situations is needed. This can be provided by responsive, flexible 'action research'. Just as we saw in Chapter 4 that ordinary people have a role in famine early warning systems, prostitutes must become AIDS educators (Schoepf 1992), and villagers must serve as water and sanitation engineers (A. White 1981).

The 'bottom-up' success in countries like China of the implementation of primary health care (PHC) is possibly more relevant than 'top-down' successes. China was able to improve health significantly through health education campaigns, and by mobilizing labour for improvement of water supply, drainage, sanitation, and housing. Schistosomiasis, sexually-transmitted diseases, and tuberculosis were reduced significantly (Horn 1965; Sidel and Sidel 1982).

Policy directions

The first steps towards the reduction of the risk of biological disaster should be the extension and strengthening of primary health care (PHC) networks. This is the goal of the World Health Organization's campaign 'Health for All by the Year 2000' (Wisner 1988b, 1992a). Left to 'market forces', health

Box 5.2   AIDS in Africa

AIDS is a disease of the immune system caused by the human immuno-deficiency virus (HIV). People may be infected (HIV-positive) for many years before full AIDS (acquired immuno-deficiency syndrome) develops, and may be unaware of their status. The disease is transmitted through sexual inter-course, or by blood from an infected person being introduced to another through shared syringes, or use of contaminated blood products (including transfusions). Inadequate precautions in the sterilization of needles and surgi-cal instruments, as well as in screening blood from blood donor services all increase the risk of contracting AIDS by means other than sexual intercourse. For these reasons, people in countries with a poor medical infrastructure are at greater risk of contracting AIDS.

There are a number of co-factors which are thought to be associated with increased rates of transmission of HIV through sex. These include genital sores, other active sexually-transmitted diseases, and cervical erosion. They are partly responsible for higher rates of infection, since the virus enters the body more easily through lesions or unhealed wounds. Poor medical facilities means that these conditions remain untreated and increase the rates of infection of AIDS. Once an individual becomes HIV-positive, the progress of the disease is profoundly affected by the pre-existing status of the immune system. Those with a damaged or stressed immune system may develop symptoms from opportunistic infections earlier than otherwise (Packard and Epstein 1987).

Endemic malaria, filariasis, war, and refugee movements have affected Uganda and neighbouring parts of Tanzania, Kenya, Rwanda, Burundi, and Zaire. Some of these areas have also been affected by human trypanosomiasis (Langlands 1968; Forde 1972; Wisner 1976a). Disrupted livelihoods, especially in Uganda, have resulted in chronic ill-health, including parasitic diseases such as malaria which stress the immune system. Typically, at least one-third of the children suffer chronic malnutrition (UNICEF 1985; Wisner 1988b).

Using a vulnerability framework to analyse the African AIDS pandemic suggests a number of social shifts that have encouraged the spread of AIDS. War, economic crisis, and upheaval of family life in Uganda have led to a greater spatial mixing of populations, and a relaxation in men's sexual control over women. *Magendo* (smuggling, literally the 'pilgrimage of greed') was rife and involved the movement of illicit goods from Mombasa, via Lake Victoria, to Rwanda and Zaire. Roving bands of traders, often armed, were away from home for months at a time. Spare cash was spent on casual sex. Indeed both smuggling along roads and across Lake Victoria, together with legal motorized traffic, gave rise to overnight stopovers and hotels, and a rise in prostitution. Many of the women (as in Buganda) have a precarious economic existence, often being *de facto* barred from owning or renting land. Thus marriage or temporary liaisons with men provided the only 'bread-ticket'. Selling beer and occasional prostitution offered ready cash in an economic climate which offered little in the way of permanent and independent livelihoods for un-attached women.

The unstable economic conditions of the 1970s added to women's deteriorating economic security, and undoubtedly led to a rapid spread of the virus. However, economic insecurity alone is not a sufficient factor explaining the cause of the AIDS pandemic. Every society gets an AIDS epidemic that also

Box 5.2 *continued*

reflects the sexual practices of its population. Behind the structuring of this pattern of sexual practice lie relations of gender inequality (Barnett and Blaikie 1992).

It is widely thought that the earlier stages of the pandemic predominantly affected urban elites in Africa, or the most wealthy with the best access to medical care. Indeed, in the late 1980s in some countries (e.g. the Central African Republic) the deaths were noticeable among 'Politicians, Professors, and Police Chiefs' (*Sunday Times*, 1 July 1990). Therefore, although economic and social insecurity are increasing the risks of becoming HIV-positive for many, this must also be understood in specific contexts of sexual behaviour that may affect the better-off. However, once an individual becomes infected, a more straightforward interpretation of vulnerability may be given. Life expectancy from initial infection is much shorter in Africa because of the greater exposure to opportunistic infections, and the lack (and high cost) of palliative medical care.

A further complicating factor has been the reluctance of some governments to evaluate or publicize the extent of infection by HIV and the number of confirmed AIDS cases. By June 1988 African governments had officially reported a total of just over 11,000 cases of AIDS. Bearing in mind that there is likely to be serious under-reporting because of Africa's weak health infrastructure, the total for the whole of Sub-Saharan Africa was still only a fraction of the nearly 66,000 cases reported in the USA for the same period (PANOS 1989). Uganda, Tanzania, Kenya, Burundi, and Congo had all reported more than 1,000 cases, and, by 1991, Uganda had reported over 17,000 cases and Kenya 12,000. Clinical AIDS prevalence was highest in the Congo, Burundi, Uganda, Rwanda, and Zambia, each with more than 300 cases per million. However, French Guiana, Bermuda, Bahamas, USA, and Guadeloupe all exceeded these African countries (with the single exception of the Congo) in published case prevalence rates (Figure 5.2).

On the other hand, aside from the possibility of unintentional under-reporting in Africa, the large numbers of infected people in selected African populations (urban prostitutes, university students, etc.) gives cause for grave alarm. There may be millions of HIV-infected young adults and children who are asymptomatic, the adults still capable of transmitting the infection, and both groups highly likely to fall ill from AIDS and related conditions within a few years.

Recently, a number of African countries have publicized data on HIV-positivity, and these are profoundly worrying. For example, it is currently estimated that 1.3 million people in Uganda (of a total population of 17 million) are infected, while in Zimbabwe up to 26 per cent of the sexually-active population are estimated to be HIV-positive (*Africa Analysis*, 17 May 1991). Other surveys of more limited but representative samples in central African countries unfortunately corroborate this gloomy forecast.

And the future? It is too early to say whether projections based on HIV-positive blood tests in urban populations are representative for the population as a whole. Nor can we say that the development from being HIV-positive to clinical AIDS (ten to fifteen years on average in the USA and

Box 5.2 continued

perhaps five to eight years in Africa) will be the same as observed elsewhere. However, assuming the worst, the effect of selective mortality could have a severe impact on much of the continent. Some aspects of this can be illustrated by using our 'access' model, as in Figure 5.3. Household labour power, livelihood options, remitted income, household income levels, purchases, and levels of welfare will all be affected. In the diagram box 2 now has a new input derived from the selective mortality on adults, which is reducing the number of active people in households. The sick need nursing, and resources are consumed in this for caring and medicine. This, combined with the reduced labour force, means that less food can be grown and earning opportunities are fewer (box 3). In some areas the deaths of parents have led to a crisis of caring for orphans, with displaced and asset-less children growing into a world in which their livelihoods are disrupted. This shows another side to the AIDS disaster that goes far beyond its medical effects.

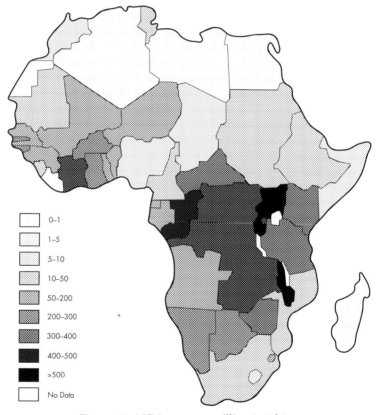

Figure 5.2 AIDS cases per million in Africa

Source: WHO Report 1992, 7.1.

services are not accessible to the poor. In fact, observing the distribution of care, Hart (1971) proposed an 'inverse care law' which states that the greater the need for care, the less is provided. Non-governmental organizations (NGOs), especially grass-roots people's development associations, can improve access to health care in a variety of ways even for the poorest households. The effects of such work in promoting safer conditions (reducing vulnerability) could be traced in several of the 'boxes' in our access model.[12]

Secondly, vulnerable groups should receive special help in improving their nutritional status. Good nutrition provides resilience. But the steps needed to improve nutrition may involve measures that are difficult to implement. Land reform, rural credit to women, access to livelihoods in the urban 'informal' sector, food pricing policy, targeted nutritional supplements, and urban gardening have all proven effective in providing rapidly improved nutritional status (Berg 1988; Biswas and Pinstrup-Anderson 1985; Cornia, Jolly, and Stewart 1987; Pinstrup-Anderson 1988).

Thirdly, agricultural research and extension should concentrate on genetic diversity, identifying and preserving local genetic diversity and resource management techniques. Integrated pest management should substitute for the import-intensive agrochemical approach to plant and animal hygiene.

Lastly, health education and agricultural extension should pay much more attention to the real constraints on people's lives. In this way, the 'messages' of educators will be more relevant: 'educators' should be more open to what they can learn from ordinary people (Barth-Eide 1978; Turner and Ingle 1985; Wisner 1987a).

Precautionary science

In the last two centuries, biodiversity has suffered irreparable damage. Our consumption of fossil fuels has begun to change the earth's climate, with a whole series of consequences for food security and health. Destruction of the ozone layer in the atmosphere could produce as many as 150 million new cases of skin cancer with over 3 million deaths in the US alone (Benedick 1991: 21). The effect of additional ultraviolet radiation on phytoplankton, zooplankton, and other life forms at the bottom of the earth's food chain is not known. In addition, large doses of ultraviolet radiation may harm the immune systems of higher animals, including humans. Such a possibility brings us around full circle, as this chapter began with a discussion of human biological adaptation and immunity.

In view of the massive changes urban–industrial civilization has brought about in a short span of time, current enthusiasm for biotechnology should be carefully reconsidered. A healthy scepticism and cautious attitude would seem appropriate before the next round of humankind's heroic environmental modification gets too far. No one foresaw the health consequences of

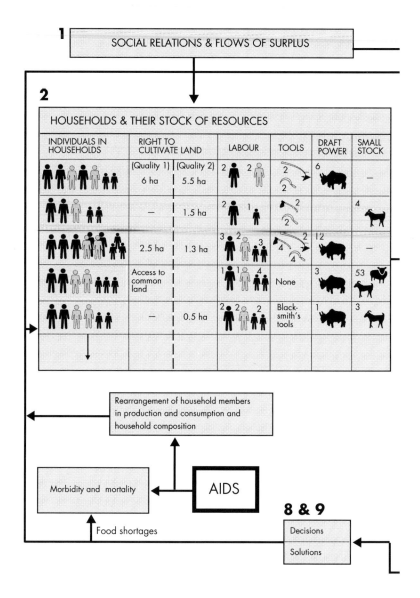

Figure 5.3 Access to resources to maintain livelihoods: the impact of AIDS

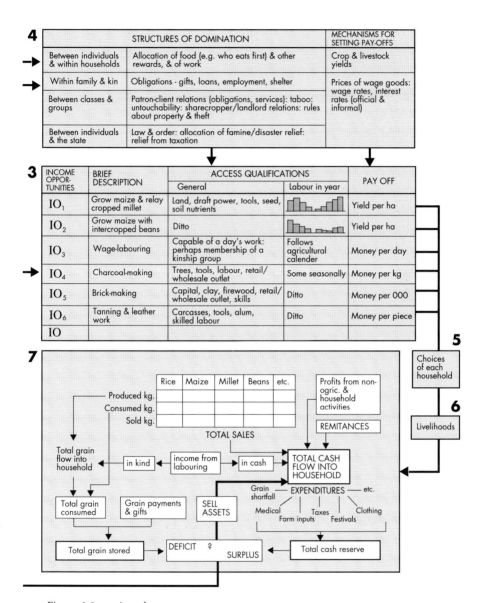

4

STRUCTURES OF DOMINATION		MECHANISMS FOR SETTING PAY-OFFS
Between individuals & within households	Allocation of food (e.g. who eats first) & other rewards, & of work	Crop & livestock yields
Within family & kin	Obligations - gifts, loans, employment, shelter	Prices of wage goods: wage rates, interest rates (official & informal)
Between classes & groups	Patron-client relations (obligations, services): taboo: untouchability: sharecropper/landlord relations: rules about property & theft	
Between individuals & the state	Law & order: allocation of famine/disaster relief: relief from taxation	

3

INCOME OPPOR-TUNITIES	BRIEF DESCRIPTION	ACCESS QUALIFICATIONS		PAY OFF
		General	Labour in year	
IO₁	Grow maize & relay cropped millet	Land, draft power, tools, seed, soil nutrients		Yield per ha
IO₂	Grow maize with intercropped beans	Ditto		Yield per ha
IO₃	Wage-labouring	Capable of a day's work: perhaps membership of a kinship group	Follows agricultural calender	Money per day
IO₄	Charcoal-making	Trees, tools, labour, retail/wholesale outlet	Some seasonally	Money per kg
IO₅	Brick-making	Capital, clay, firewood, retail/wholesale outlet, skills	Ditto	Money per 000
IO₆	Tanning & leather work	Carcasses, tools, alum, skilled labour	Ditto	Money per piece
IO				

5 Choices of each household

6 Livelihoods

7

	Rice	Maize	Millet	Beans	etc.
Produced kg.					
Consumed kg.					
Sold kg.					

TOTAL SALES

Profits from non-ogric. & household activities

REMITANCES

Total grain flow into household — in kind — income from labouring — in cash — TOTAL CASH FLOW INTO HOUSEHOLD

Total grain consumed · Grain payments & gifts · SELL ASSETS

EXPENDITURES — etc.

Grain shortfall · Medical · Taxes · Clothing · Farm inputs · Festivals

Total grain stored → DEFICIT ? SURPLUS ← Total cash reserve

Figure 5.3 continued

releasing tens of thousands of newly synthesized chemicals into the environment until Rachel Carson wrote *Silent Spring* (1962). What will the deliberate release (and accidental escape) of genetically engineered life-forms do? There are certainly some beneficial uses of recombinant DNA technology in medicine and agriculture. But who should decide which application is worth the environmental risks (Walgate 1990)? If we have learned anything about human health since Hippocrates wrote *Airs, Waters and Places*, it is that each person's health is inseparable from the health of others and the health of the environment. Reduction of vulnerability to biological hazard will therefore require both social justice (to ensure the health of others) and technological humility (to restore the health of the environment).

NOTES

1 On the epidemiology of disasters, see Chen (1973); De Ville and Lechat (1976); PAHO (1982); UNDRO (1982b); Seaman, Leivesley, and Hogg (1984); Alexander (1985); Sapir and Lechat (1986).
2 For analysis of African migrants who contract malaria see Prothero (1965). These arguments have also been advanced in the health sciences. See discussions of 'political epidemiology' in Brownlea (1981); Doyal (1981); Packard (1989); and Turshen (1989).
3 Here we would include exposure of the migrant to sexually transmitted-diseases (STDs) – possibly including AIDS – and tuberculosis. These are known to have been spread to rural areas which provide contract labour to the South African mines, for instance (De Beer 1986; Packard and Epstein 1987). The spread of HIV infection from urban centres to villages is also documented (Barnett and Blaikie 1992).
4 Dutch agricultural production fell by half from 1938–44/5 due to disruption caused by the war. Nevertheless the Nazis first thoroughly integrated the Dutch economy into their war efforts, exporting large quantities of food to Germany. Then, as a reprisal for the Dutch railroad strike, they cut off all imports of food, fuel, and electricity into the Netherlands during the last eight months of the war. Finally, the occupiers intentionally flooded 8 per cent of the country to impede the Allied advance, further weakening agricultural production. The result was famine, with urban rations as low as 500 Kcal in January 1945 (one-third of the adult subsistence level). Twenty soup kitchens were run by Amsterdam city officials, feeding up to 160,000 people daily. People subsisted on beet-sugar and food obtained from the black market, where a loaf of bread could cost US$27. In view of the severity of the famine it is remarkable that only 15–18,000 Dutch perished out of a population of about 9 million (Mass 1970; Warmbrunn 1972).
5 Some information on health crises is given in Chen (1973) for the Bangladesh war of 1970, and in CIMADE, INODEP, and MINK (1986) and Kibreab (1985) on refugees in Africa.
6 The impact often equalled or surpassed that of the Black Death in Europe. Thomas Jefferson, for instance, wrote in 1781 of the Indians of the Powhatan confederacy in Virginia:

> What would be the melancholy sequel of their history, may however be augured from the census of 1669; by which we discover that the tribes

therein enumerated were, in the space of 62 years, reduced to about one-third of their former numbers. Spirituous liquors, the small-pox, war, and an abridgement of territory, to a people who lived principally on the spontaneous productions of nature, had committed terrible havoc among them.

(Quoted in Peterson 1977: 135)

McNeil (1979) also records the devastating impact of smallpox in 1530–1 on the Indian ability to withstand the military invasion of the conquistadors.

7 Major pandemics of bubonic plague occurred in the mid-sixth to mid-seventh centuries (Plague of Justinian and its aftermath in England and Ireland), the latter half of the fourteenth century (Black Death), and mid-seventeenth to mid-eighteenth centuries, including the Great Plague of London in 1665 (Marks and Beatty 1976).

8 Extensive sewer systems were first dug in Hamburg in 1844–88 and London in 1854–65. The association between cholera and water used for drinking and cooking was suggested only in 1849 by Dr John Snow (Read 1970).

9 The intense concentration on potato cultivation in Ireland was an anomaly compared with accounts of peasant farming in other parts of Europe in the same period (Shanin 1971).

10 By a conservative count, more than 4 million people have died directly or indirectly in war, the largest totals being: more than 2 million in Biafra (Nigeria), 600,000 in Ethiopia, 550,000 in Uganda, 500,000 in Mozambique, 300,000 in Sudan, and more than 100,000 each in Zaire, Burundi, Rwanda, and Angola (Barnaby 1988).

11 Of course there is the question of what these large quantities of insecticide may do in the environment, but in general this book cannot deal with the enormous area of technological hazards. See Weir (1987) for an account of the Bhopal disaster in India.

12 Some NGOs have challenged primary health care as 'not cost-effective'. This challenge arises out of the debate that has accompanied a retreat from 'basic needs' as a fundamental goal (Wisner 1988a; Newell 1988).

6

FLOODS

INTRODUCTION

Perhaps surprisingly, flooding is considered by some authorities to be the hazard that affects more people than any other (Ward 1978; UNDRO 1978: 1). Floods may even affect places which at other times are prone to drought. Yet, in many parts of the world, floods are also an essential component of the social and ecological systems, providing the basis for the regeneration of plant and aquatic life and of livelihoods derived from them. Some people (e.g. in Bangladesh) have different terms to distinguish between beneficent and destructive floods. This ambivalent character of floods is important, and is discussed later in the subsection on livelihood disruption.

Flood disasters are destructive of life not only through drowning and direct injury, but also because of associated diseases and famine. Their impact must also be measured in terms of the disruption and destruction they cause to livelihoods, and the changes in the access profiles of affected peoples. The loss of assets or ability to work, or of land and animals, and the consequences of injury and illness may be felt for many months or even years after the inundation has subsided. Any deaths which occur after such a time-lag are unlikely to be linked officially with the flood. As with famine and biological disasters, the access model indicates how vulnerability to *future* floods (or other hazards) can be increased by the longer-term impacts of floods on household assets, labour power, and social networks.

Human activity and livelihoods lead to people being located in rural and urban areas which are prone to flooding, and the number of people vulnerable is increasing as populations rise and the lack of alternative settlement sites places increasing numbers on flood plains. So vulnerability to floods is partly a product of human-created environments, though the risks are experienced in varying degrees by different groups of people. Economic and social systems allocate societies' resources to the detriment of some groups and the benefit of others, and this affects people's capacity to withstand floods, and also exposes them to flood risks unequally.

For instance, new flood hazards in towns and cities may be generated by

124

economic and social pressures that force marginalized people into flood-prone urban locations. Soil is covered by impermeable surfaces, and settlements which encroach on hillsides increase the rate of runoff, leading to floods where there were none before. C. Green (1990) emphasizes that, because such urbanized gulley-type catchment areas are generally small, the floods tend to be 'flash-floods', and are therefore likely to produce high mortality. He argues that as urbanization continues very rapidly in the Third World, the balance of flood risk is likely to shift from rural to urban areas. (See also Chapter 8 on landslides, which are often associated with floods and affect shanty-town dwellers.)

Flooding may also be associated with famine (as, for example, in Bangladesh in 1974), and in such situations famine may be the more significant cause of death. Even where flood disaster does not extend into famine, the impact of the deluge on many people's livelihoods is at least medium-term disruption and probably means hunger for some groups of people. Land and other assets may be lost or have to be sold. Usually long-term vulnerability is increased, so that starvation is a more likely outcome from the next hazard-strike, whatever the biological or geophysical process involved (e.g. cyclone, drought, earthquake, plant pest attack). For instance, if the next hazard reduces wage-earning opportunities for agricultural labourers (perhaps if there is no harvest to weed or gather), the associated higher food prices in the markets may lead to starvation for some people even when there are adequate food supplies (Crow 1984).[1] Our discussion of Sen's theory of exchange entitlements (of FAD versus FED) in Chapter 4 explained why this is a possible outcome.

FLOODS AND KNOWN RISKS

As a result of prior flooding, the places affected by flood hazards are generally known. However, this is complicated by the wide ranges of intensity of floods in the same areas, and the variation in return periods (the average number of years' gap between floods of a given magnitude recurring). Areas that are at risk from floods often offer considerable benefits (for farming or industry), as do the fertile slopes of volcanoes (see Chapter 8). The gains in farming include farm use of flood-plain alluvium (usually with better soil and water moisture retention), and the land-cost advantages of industrial and urban locations on flood-plain sites. Less often considered are the enhanced fishing opportunities derived from the nutrient-rich waters brought to ponds, lakes, and rivers by freshwater inundation. In this sense, human action through settlement patterns has created the flood risks, though there are many variations in the degree of vulnerability of different sections of a population.

There are also floods which can be almost entirely attributed to human action. The hazard may arise downstream as a result of the supposed benefits to livelihoods from economic activity in upstream locations, sometimes a considerable distance from the flood-zone itself. The most prominent examples of such problems are of dams built (whether or not to control floods) to inadequate standards, or on unsafe sites, which collapse or lead to other failures. This causes often devastating flash-floods (such as the over-topping of the Vaiont dam in Italy in 1963, with up to 2,000 killed). But it is also worth recognizing that some dams shift the flood-water problem elsewhere, as with the Ganges' peak flows diverted through Bangladesh by India's Farakka Barrage (Monan 1989: 27). Other dams built for hydro-electric power and irrigation (or even flood control) often result in permanent flooding of inhabited areas, as with the large dams on the Narmada in India, which will forcibly move up to a million people.[2]

DISASTROUS OUTCOMES FOR VULNERABLE PEOPLE

Understanding how flood disasters happen requires analysis of the various patterns of vulnerability generated by different economic and political circumstances. These may then be linked to the bundle of factors which make up the access profile of different people that determines their level of vulnerability. This should enable the identification of the people at risk in flood-prone areas.

Mortality, morbidity, and injury

Floods are not only one of the most widespread of natural hazards; they also lead to the greatest loss of life, immediately through drowning and fatal injury and through illness and famine.[3] Others may die later, never to be counted as victims of the deluge, because their livelihoods suffer a further downward twist of the spiral.

In a wide range of Third World countries, floods frequently lead to large numbers of deaths. Flash-floods are particularly hazardous, because in many places there are people in locations where this risk is not offset by pre-cautions or warning systems. For example, in 1988 nearly 600 people were killed or reported missing following flash-flooding in China's coastal province of Zhejiang (*Guardian*, 3 August 1988). Extreme events with long or unknown return periods are understandably difficult to anticipate. But it ought to be possible to take precautions in many upland or hilly regions of the world where the risk is of flash-floods with much shorter return periods.

Combination floods, in which there is a coincidence of riverine inunda-tions with heavy rainfall or coastal storms, are much more disastrous than ordinary slow-onset floods. Aside from the many casualties which can be caused by flooding brought about by the impact of tropical cyclones, the highest direct mortality figures appear to result from other rapid-onset deluges. These may result from tsunamis (so-called tidal waves, caused by undersea landslides, volcanic, or earthquake activity) as well as flash-floods. Slow-onset riverine inundations of flood plains result in lower direct casualties. These casualties are more often from building collapse (people may remain in upper storeys or on roofs), other injuries, and snake bites than from drowning. Slow-onset floods also enhance the risks of disease and malnutrition in the months or even years following.

This pattern is also recognizable in combination (riverine and rain-fed) floods like those which led to the 1988 Khartoum (Sudan) disaster. The people most severely affected by these floods were the millions who were already refugees from famine and the civil war in the south of the country. Many had settled in low-lying, flood-prone land around the capital city of Khartoum (Abu Sin and Davies 1991a; Woodruff *et al.* 1990), where they were highly vulnerable to epidemic disease and (as apparent in 1988) to flooding. By 1992 the authorities had decided to move hundreds of thousands of these squatters away from the capital city. Early reports from international aid workers suggested that a number of people had been killed by soldiers in the process of forced removals and worries were expressed about the availability of shelter, food, and water at the new sites in the desert (Crossette 1992).[4]

Bangladesh experienced immense slow-onset flooding (much of it from local rainfall, as well as rising rivers) for two years running in 1987 and 1988. These came on top of the devastation of 1984 and 1974, disasters which had increased many people's vulnerability (see Box 6.1 on Bangladesh at the end of this chapter).

Flood-waters bring increased risk of disease such as cholera and dysentery, arising from sewage spread and the contamination of drinking-water. There may be rapid growth in the incidence of malaria and yellow fever because of the multiplication of insect vectors in the stagnant water, which may remain lying about for months after an inundation. In some regions (this is common in Bangladesh and India), water is held back by raised structures like roads and railways, and these cause local floods that are sometimes severe. These structures should have ducts and culverts which permit the return flow of water back to river channels, but often these are inadequate or badly maintained. In some floods, they may also be blocked by the large quantities of silt that are deposited.

Respiratory illnesses often become more prevalent in the aftermath of slow-onset floods, and take a toll especially among very young children and babies and the elderly. Illness or injuries caused in the flood are important factors which increase existing vulnerability, and extend vulnerability to new groups of people. The sick and injured usually cannot work, and the family's loss of their labour, especially during attempts to recover after a hazard-strike, can be an element of the disaster.

There are few surveys of what actually happens to people after floods strike. An idea of the pervasiveness of morbidity and disablement problems is given by a 1980 sample survey in Pakistan of rural settlements in the Ravi valley. The people interviewed were asked about their experiences in three years in the previous decade in which floods were particularly bad. Of the families questioned, between 43 and 57 per cent of members fell ill after floods, and 'at least one member of every family [was] bed-ridden throughout the coming season' (Sikander 1983: 102). Pakistan regularly experiences flooding affecting around 700,000 people a year, though in bad years (like those of 1971, 1975, and 1979 investigated in the survey) between 3 and 6 million have been affected.

The health problems are particularly highlighted in studies of floods on the west coast of South America brought by El Nino in 1982–3.[5] In that southern summer, El Nino struck badly, principally affecting Peru and Ecuador. In parts of Peru a state of emergency was declared: rainfall in the first six months of 1983 was many times more than the total rainfall of the previous ten years (Gueri, Gonzalez, and Morin 1986). Flash-flooding and landslides destroyed many roads, irrigation facilities, dams, and bridges. Mortality directly caused by floods does not seem to have been high, but disease and health problems were made much worse, and people's livelihoods suffered enormously.

In Ecuador, many rural people, faced with the failure of waters to subside, fled to towns and cities, more optimistic about conditions there. They brought malaria infection with them, leading to the reinfection of urban areas previously cleared of the disease. The floods greatly increased the number of cases of malaria anyway. Despite massively increased insecticide spraying, the number of cases rose in 1983, and even more in 1984, to levels ten to twenty times (depending on location) those of previous years (Cedeno 1986).

To the south, in neighbouring north Peru, a study of government health centres showed morbidity rates up by 75 per cent for respiratory and 150 per cent for gastro-intestinal illnesses in the first six months of 1983 (compared with the same period the previous year when El Nino had not been extreme) (Gueri, Gonzalez, and Morin 1986). These illnesses led to a large increase in death rates. The centres surveyed covered a population of 630,000. The number of deaths in the first half of 1983 was 6,327 compared with 3,226 in the same period in 1982.

Livelihood disruption

While death, illness, and disablement lead to a reduced capacity for work in affected families, there are other impacts on people's livelihoods which make some more vulnerable and enrich others. Not all groups in flood areas are necessarily disaster victims. The flood may have its impact on different social and economic groups in a more or less severe manner. In floods it is of course true that much property is damaged, destroyed, or swept away. But even flooded land can be sold by a destitute farming family to buy food, despite its likely low price arising from the many similar 'distress sales'. The same applies to other goods, and so there are beneficiaries of the disaster who can accumulate land or other assets at depressed prices. Others may benefit from their possession of food stocks, selling at higher prices in the aftermath. Still others may have saleable goods or services on which they can thrive, perhaps trading in drinking-water by virtue of owning a boat to carry it around.

Each household's 'bundle' of property and assets (including land and animals for farmers, or boats and nets for fishermen) and economic connections with others may be lost, enhanced, disrupted, or reinforced in a number of permutations. This sort of disaggregating approach to the impact of hazards shows that although possibly a large majority of people are made worse off, floods may not be a disaster for everyone. They operate under the influence of rules and structures derived from the existing social and economic system, but modified by the distinct characteristics of particular flood strikes and patterns of vulnerability.

People in Third World countries are rarely insured. The loss of the home is a major livelihood set-back, not because it is necessary for earning a living (though it often is), but because of the burden on limited finances in providing some replacement. This cost may not be in terms of cash outlay, but instead the loss of time which would otherwise be used in livelihood/earning activities. In Chapter 3 we provided one striking example of the difference between rich and poor in rebuilding houses following a cyclone and flooding in Andhra Pradesh, India.

Many simple household items may also need replacing, such as cooking pots and water vessels. This also diverts time and labour from livelihood activities, or consumes limited reserves. Uninsured people with no reserves of cash lose twice in a flood disaster: they lose the goods, many of which are essential to life, and they lose the time which they have to spend in work to replace them, which is therefore not available for survival (in food-growing or wage-earning). Having reserves or insurance means being able to return more immediately to normal livelihood activities.

Other losses may be directly disruptive of household livelihood. Standing crops are a loss to the farmers who own them (and for the poorer families this is perhaps the most serious aspect of flooding). In many areas of the world, there is an unhappy coincidence because the season in which floods are more likely is also the one in which the crops ripen for harvest. An added 'ratchet effect' arises because this pre-harvest season is also often the 'hungry season' when household food stores and income are low and the physiological reserves of people are depleted (Chambers, Longhurst, and Pacey 1981).

Crops in some parts of the world are well-adapted to expected levels of flood. Many thousands of indigenous varieties of rice have been developed in South and Southeast Asia. They include types which grow so rapidly they can keep pace with rising flood-water, and the floating varieties planted in many areas, selected to be grown with floods. 'It is estimated that over 20,000 varieties of rice have been developed by farmers to suit the different cropping conditions in Bangladesh' (Oxfam 1989). But even these will succumb to inundation under some circumstances, along with non-adapted crops.

Larger landowners do not need labourers in fields when these are flooded. The consequent loss of wage-earning employment may be disastrous for those families which rely for a large part of their livelihood on such income-earning opportunities. This was what caused great hardship in Bangladesh following the 1974 floods (E. Clay 1985). The impact of crop loss for the better-off landowners is likely to be much less disastrous, depending on the amount of the families' reserves.

The length of time that water remains on the land may also affect the prospects of the subsequent normal planting, or of a 'catch crop' aimed at recovering some of the losses. For example, in Bangladesh there is often a good harvest associated with floods. In non-flooded areas, there is often a coincidence of good rainfall which increases yields. In flooded areas, the soil retains more moisture which can be taken up in plant growth during the dry season. In 1988, even after the worst floods ever experienced in Bangladesh, the harvest was record-breaking. However, a household whose water buffalo has died in a flood, or which for other reasons cannot take advantage of these soil moisture conditions, will not be able to plant in a timely way and 'catch up'.

Such animals may be swept away and drowned or injured, and their loss to those families which use livestock produce for subsistence or sale is comparable to those reliant on crops. Animals are often the main source of draught power and transport for significant sectors of the rural population in many parts of the Third World. In Sikander's (1983) study of Pakistan, the surveyed villagers reported 35 per cent losses of their animals. The risk of the death or injury of these crucial animals in floods adds a further measure of vulnerability.

Recovery is often not to the same level of well-being as before the hazard struck. Crops may be grown again (if the soil is undamaged) within six months, whereas large livestock is a considerable investment for many people, and it may take years to recover from losses. Social position and esteem may decline to the detriment of some people. A study in Botswana reported that the status of the elderly declined after a tornado destroyed a quarter of the houses in an area, followed by torrential rain that flooded the harvest. In the aftermath the knowledge and resources of the elderly were important, but during long-term reconstruction it was the middle-aged group that commanded government resources. Whereas the elderly had lived independently in their own houses before the disaster, they became dependent afterwards (Guillette 1992).

In floods there is a physical process by which land is destroyed by the erosive capacity of the flood streams, and recreated in the areas where silt is deposited as sediment-laden waters are slowed down. Flooded rivers are by definition flowing beyond their usual banks. Their route across the countryside, if unconstrained by human constructions, will be through new routes provided by the lowest-lying land. Rivers carve new channels in this way, often kilometres away from their previous course.

Those whose land is lost in this process are unlikely to have access to the compensation in the form of land to replace it, even should they get other types of aid. Yet others may find that, fortuitously, the river has abandoned a channel near them, making it possible in time to colonize the waterlogged land. However, the more powerful and already better-off households are more likely to gain control of such new land, as happens in Bangladesh (Elahi 1989).

Land is 'lost' in other ways too. Depending on the speed of the flood-waters at a given place, the soil itself may have been carried away. Generally, though, as flood-waters spread out across the landscape, they slow to a pace at which they can no longer carry their suspended load of silt and sand. They then deposit the sediment on top of the earth. In some regions it is usually beneficial in replenishing minerals which are useful to plant growth and otherwise improving fertility. The nature of the silt deposited is benign in such a case, but this is not always so. The size of deposited particles may be much larger, covering extensive areas with infertile sand or gravel. The mineral content of the sediment may be too saline or alkaline, rendering the ground toxic to plants.

Depending on the combination of these different factors, the land left behind can be enriched and newly fertilized by the layer of deposited silt, or made more barren and less productive, with a crust of inferior sand or minerals which inhibit plant growth. Flash-floods in Rajastan (west India) are likely to produce the latter situation (Seth, Das, and Gupta 1981) in a region which is more usually facing up to problems of drought rather than inundation. The Kosi's floods in north Bihar also normally deposit a layer of sand over agricultural land, rendering it useless for up to fifty years (Lyngdoh 1988).

In Bangladesh it is more likely that flooded land is enriched by the new layer of silt cast on it by the floods. Further downstream in that country, the waters of the Ganges and Brahmaputra arrive at the Bay of Bengal laden with silt, are stilled by the sea and add new material for the expansion of the delta. This new land, often in the form of islands called *chars* in the middle of the many channels of this complex river system, is quickly occupied by poor and landless peasants from elsewhere who otherwise have no means of subsistence (Zaman 1991). Their precarious existence on the edge of this watery boundary-zone is discussed in Chapter 7 on cyclones and coastal flooding.

Water itself is an important part of the resource or livelihood rights of many people likely to be affected by floods. Rivers are crucial for livelihoods based not just on agriculture but also on transport, trading, and fishing. Shifting river channels may disrupt these livelihoods too, creating havoc among whole sections of a population.

In some circumstances, the normal flood regime of a river is used beneficially by farmers. On some rivers 'flood-retreat' agriculture (also known as recession agriculture) is practised where the receding waters reveal moist soil primed for planting with food crops. It used to happen on the Nile before the Aswan dam was installed. Such a system has existed for centuries on many rivers in Africa, including the Senegal on its route through Mali, Mauritania, and Senegal in West Africa, as well as in South Asia.

Livelihoods based on farming and fishing (in the ponds which remain as the flood goes down) are severely undermined when 'development' projects attempt to control such a river. For example, a dam has been constructed at Manantali on the upper Senegal, mainly for the generation of hydroelectric power and to regulate the river's flow to permit year-round barge traffic up to Kayes in Mali. It will also irrigate farmland, but in large-scale projects which will grow crops local people do not eat. This will not compensate those who lose out in the transfer of resources and as a result of the loss of natural floods in valley.

One of the costs of such projects is that the new regulated river flow will not allow traditional flood-retreat agriculture to take place (Horowitz 1989; Horowitz and Salem-Murdock 1990). Similar losses to small-scale traditional irrigation were the result of damming a river to the south of Kano, Nigeria so that large-scale irrigated wheat production could supply urban bakeries (Andrae and Beckman 1985). Other examples of loss of prior agriculture and fishing resources in African rivers with flood-recession capability are given in Fiselier (1990) and T. Scudder (1989).

FLOODS AND VULNERABILITY

Flood hazards impart a variable impact on people according to vulnerability patterns generated by the socio-economic system they live in. Those who are vulnerable to a hazard are unlikely to be able to move against the process which has generated their vulnerability, so that after a hazard's impact they are yet more vulnerable to similar and other hazards.

Class relations and structures of domination are crucial for explaining vulnerability to floods. They determine levels of ownership and control over assets and means of production, together with the resultant livelihood opportunities which may be already inadequate to provide basic needs indicated in the household budget.

132

Our access model thus explains many specific mechanisms that turn flood hazard into flood disaster. These include the location of homes (and their proneness to inundation) and the structure and type of housing and workplace (and their resistance to floods). Both of these are a function of household income, legal or social limitations on land-use, availability or cost of building materials, and the location of livelihood activities. The access model also describes the daily and yearly pattern of work and other activities. These in turn interact with the temporal patterns of flood hazard occurrence. These variables not only affect the risk of death and injury but also the risk of destruction of assets and livelihood opportunities. All these can be summarized in the 'pressure' model for floods, which shows how the more remote root causes of flood vulnerability are translated into unsafe conditions through the action of various pressures (Figure 6.1).

Ethnic divisions are often superimposed on class patterns, or in some situations become the dominant factor structuring vulnerability. This involves differential access to or possession of resources, or inequalities in participation in different livelihoods, according to imposed racial or ethnic distinctions. For example, the impact of exceptional flooding around Alice Springs in central Australia in 1985 was felt more by Aboriginal people, who did not receive flood warnings and who lived in flimsy accommodation on low-lying land. The radio broadcasts that alerted the white people were not on channels which were customarily used by the Aborigines (*Hazards Panel Newsletter*, November 1985).

It is also crucial to understand differential vulnerability dependent on gender. Many of the material differences between classes mentioned in the section on livelihood are comparable with women's and men's unequal possession of, and access to resources. In general terms, economic and cultural systems are male-dominated, and allocate power and resources in favour of men. In relation to flood hazards, this may mean that the efforts put into disaster recovery are disproportionately carried by women, who in most 'normal' situations have to work harder in rural agricultural and domestic activities.

In addition, there is the possibility that women are likely to be more prone to post-flood disease, largely as a result of their poorer nutritional condition and physical susceptibility. Men's and women's time and place patterns of daily and seasonal activities also differ, and this may produce inequalities in their exposure to flood hazards. To the extent that young children are more likely to be with women than men, this also affects their relative vulnerability.

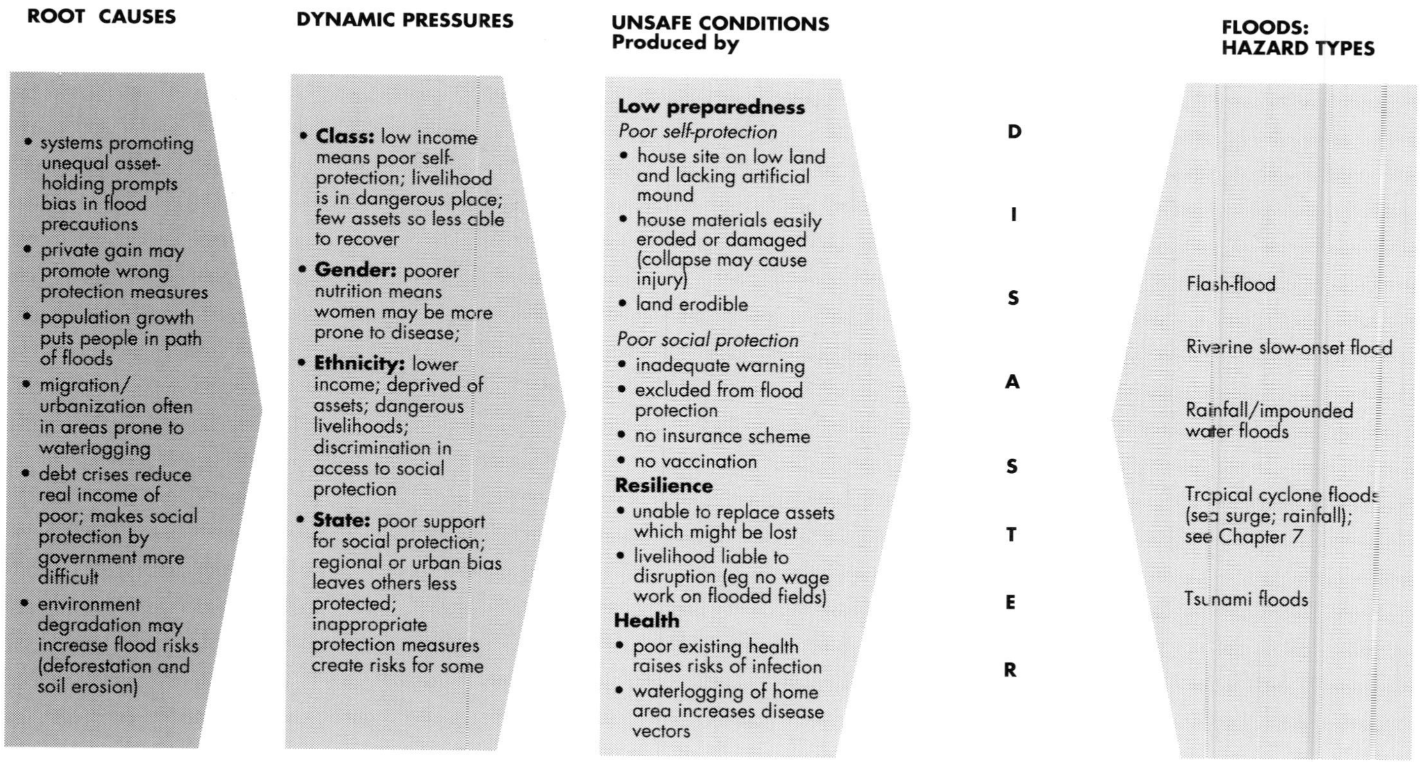

ROOT CAUSES

- systems promoting unequal asset-holding prompts bias in flood precautions
- private gain may promote wrong protection measures
- population growth puts people in path of floods
- migration/ urbanization often in areas prone to waterlogging
- debt crises reduce real income of poor; makes social protection by government more difficult
- environment degradation may increase flood risks (deforestation and soil erosion)

DYNAMIC PRESSURES

- **Class:** low income means poor self-protection; livelihood is in dangerous place; few assets so less able to recover
- **Gender:** poorer nutrition means women may be more prone to disease;
- **Ethnicity:** lower income; deprived of assets; dangerous livelihoods; discrimination in access to social protection
- **State:** poor support for social protection; regional or urban bias leaves others less protected; inappropriate protection measures create risks for some

UNSAFE CONDITIONS
Produced by

Low preparedness

Poor self-protection
- house site on low land and lacking artificial mound
- house materials easily eroded or damaged (collapse may cause injury)
- land erodible

Poor social protection
- inadequate warning
- excluded from flood protection
- no insurance scheme
- no vaccination

Resilience
- unable to replace assets which might be lost
- livelihood liable to disruption (eg no wage work on flooded fields)

Health
- poor existing health raises risks of infection
- waterlogging of home area increases disease vectors

D
I
S
A
S
T
E
R

FLOODS: HAZARD TYPES

Flash-flood

Riverine slow-onset flood

Rainfall/impounded water floods

Tropical cyclone floods (sea surge; rainfall); see Chapter 7

Tsunami floods

Figure 6.1 'Pressures' that result in disasters: flood hazards

FLOODING AND DEFORESTATION:
THE CAUSATION CONTROVERSY

There is a widespread assumption by those concerned (for very good reasons) about deforestation that flooding may often be increased as a result of land being cleared of vegetation. This causation is frequently mentioned in relation to floods in Bangladesh and north India, and is linked to deforestation in the foothills of the Himalaya. Studies of this region challenge this assumption and clarify the issue. Since it is of considerable relevance to vulnerability to flooding, it is discussed here briefly.

Land shortage in upland areas may increase the rate of deforestation, as people clear more land for agriculture, or damage trees for fuel and fodder. They may be new arrivals who clear new areas, or local people who have to expand their cultivated area or reduce the fallow period in upland swidden (slash-and-burn) agriculture. Many people associate such deforestation with a supposed increase of flooding downstream.

In upland areas of many parts of the world, a range of factors produces landslips and soil erosion. These may generate local flooding (through stream-damming) and increase the sediment load of rivers, contributing to the rise in the level of river-beds downstream and an increased flood hazard. However, there are disputes among scientists about the significance of different factors in this process, especially concerning the Himalayas. One disagreement is about whether or not there has been an increased incidence of flooding during recent decades when, it is supposed, rapid deforestation has occurred.

Some argue that the evidence for a strong connection between deforestation and increased flooding is uncertain, and that hydrological data do not demonstrate that good vegetative cover in large river basins is necessarily a factor in preventing rapid runoff of storm-water (Ross 1984: 224–5). Others suggest that flooding of equivalent severity and frequency has occurred for centuries in river basins, long before recent increases in deforestation. For example, discussing the situation in Sichuan province, Ross (1984: 223) presents arguments by one Chinese engineer that 'historical records show a high incidence of flooding even before modern increases in population and logging'. Ives and Messerli (1989) argue likewise for the Himalayas, saying that there is no convincing evidence of an increase in runoff during the last forty years, despite the apparent increased incidence of flood disasters. The rivers of the Ganges–Brahmaputra basin have been contributing immense amounts of sediment to the Ganges plain and Bengal delta for thousands of years, owing to climatic and tectonic factors in the mass wasting of Himalayan slopes, rather than recent human action. Ives and Messerli ascribe the common perception of an increase in flood disasters not to greater amounts of water in the drainage system, but to human systems having put more people in more risk-prone places.

The significance of this controversy for a discussion of vulnerability to floods is twofold. Firstly, if we are to accept that vulnerability is a condition deriving from economic and social systems, it is not certain whether the deforestation process should be included as a significant contributor to rising vulnerability. Secondly, arising from this, reducing vulnerability in areas downstream of significant deforestation may not necessarily be achieved by curtailing that deforestation (or by reforestation), although such a policy may be of major benefit in other ways.

FLOOD PREVENTION AND MITIGATION

Precautionary measures and policies for dealing with floods are aimed at modifying or predicting the hazard involved in the triggering of flood disasters, rather than with the other causes of vulnerability. They include strategies intended to reduce the intensity of the hazard, different forms of precautionary intervention, mitigation of flood effects, prediction and preparedness. However, we feel that policies need to go beyond this to look at the implications of vulnerability analysis in the development of different ways of disaster avoidance.

Local-level mitigation

Local-level, indigenous responses include people's own strategies for dealing with flood risks. These entail a combination of self-protection and social protection by communities or non-governmental agencies. These responses have been developed by people in many places, often over hundreds of years, especially where people have had to colonize and cultivate new lands in flood plains. In some regions, for instance in parts of India and in Bangladesh, rural houses are usually built on artificial mounds that raise them above normal flood levels. But this is not always possible for all people. For instance, in the Gangetic plain of north India, there is often higher ground at the centre of villages in flood-prone areas. The more substantially-built houses of the wealthier groups are often near these village centres. Poorer classes, including lower castes and untouchables, are mainly to be found round the edges of the settlement in low-lying sites.[6]

Flood prevention

One common response to riverine flood hazards is to attempt to ameliorate their disastrous impact by modifications of the stream flow. Discharge controls include a narrow range of measures which nearly always involve a high level of technical (and therefore capital) investment. Large-scale dams and barrages are sometimes used in this context. In many cases they have been successful in flood mitigation and prevention, as, for example, with the dams on the Damodar River in Bihar and West Bengal, which are claimed to have greatly reduced flood damage in 1978 (Government of India 1978: 3). On the Yellow River in China, dams in the upland tributaries have been credited with preventing serious flooding in 1981, despite an unprecedented peak discharge (*Beijing Review*, October and November 1981). These dams are still being modified and added to in order to complete the comprehensive flood protection works over the length of the river (*Beijing Review*, 18–24 June 1990).

A problem of large-scale investment projects is that they can induce a false sense of security. This may be misplaced if the design capacity of the dam is inadequate, or siltation is greater than expected. Furthermore, the dam itself may fail (collapse) because of design faults, construction inadequacies, incorrect location on inappropriate rock-base, or earthquakes (which may sometimes be locally-induced by the mass of the reservoir water itself). There have been many dam failures in industrialized countries as well as the Third World. For instance, in 1975 the Grand Teton dam in Idaho (US) failed, doing US$2 billion worth of damage in a number of downstream towns. Only eleven people died because there had been warnings and an evacuation, but 4,000 homes were destroyed along with 350 businesses. One estimate suggests that 'on average ten significant dam failures have occurred somewhere in the world each decade, in addition to damaging near-failures' (Veltrop 1990: 10).

Channel control methods often involve employing thousands of workers along lengthy stretches of river. The most common approach is to constrain river channels within artificial embankments or dykes (river training), or to use dykes (called bunds in some countries) to protect particularly vulnerable areas along the river. In addition, embankments may be used to encircle areas or places (e.g. ring bunds around towns or cities) which are deemed to need special protection. Such methods have a long history in some parts of the world, including India and China. In China the channel of the Yellow River (Huanghe) for much of its course across the North China Plain has been repeatedly enclosed within dykes for thousands of years. As a result, for some of its journey it now flows in impounded channels 5 metres (16 feet) above the level of the surrounding countryside. It is a policy that requires massive expenditure, and the dykes must constantly be raised higher and higher (*Beijing Review*, 30 July–5 August 1990).

There are other channel control methods which are used (often in conjunction with river training) to provide emergency storage for flood-water (called 'detention basins'). These may be existing lakes which adjoin the river channel, or artificial depressions or low-lying areas. The embankment leading to the lake can be deliberately breached, and water from the peak flow is then stored to prevent the river reaching danger levels further downstream.

Box 6.1   Bangladesh – A 'tech-fix' or people's needs-based approach to flooding?

Since the extensive floods of 1987 and especially 1988, Bangladesh has attracted much international concern in 'solving' the flooding problem. In 1988 the government of Bangladesh (GoB) in co-operation with the UN Development Programme (UNDP), plus official teams from Japan, France, and the United States, began studies of the flood problem (for a summary, see World Bank (1990: 25–30)). These studies differed considerably in their prescriptions for dealing with the hazard, ranging from a capital-intensive 'high-tech' inter-vention (France) to a version of a 'living-with-floods' approach (USAID).

The World Bank has taken on the co-ordination of foreign and GoB proposals. It produced a report known as the Bangladesh Flood Action Plan (BFAP) in conjunction with the GoB, which amalgamates some of the ideas from the various other studies (World Bank 1990). It describes initial spending of US$146 million in various preparatory studies, with some capital going into the repair of existing projects. The Bank has in effect become a broker between the different interests represented in the Group of Seven (G7) richest countries and GoB, trying to keep on board all potential donors, and mediating between the different flood plan proposals.

This is a difficult task, since many of the plan's components conflict with each other (both in terms of their technical requirements and their objectives). Also some proposals are of very doubtful financial viability, and would fail to be approved if judged by normal World Bank auditing criteria (see Boyce 1990: 422–3). The key issue is which approach is favoured by the BFAP supporters, especially the GoB. Will it involve high levels of investment in large-scale engineering works that will supposedly contain the rivers? If so, will this 'river training' approach also help by promoting agriculture in protected areas, including by providing dry-season irrigation? Or are the studies going to produce flexible and appropriate proposals for different areas and address the needs of the people? While the proponents of the plan believe it to be flexible and not overemphasizing a 'technical fix' (e.g. Brammer 1993), there is considerable cause for concern given the widely-based and serious criticisms emerging from many people in Bangladesh, including affected people in the countryside, academics and engineers, and many NGOs.

At present, it is not certain what major proposals for projects will arise from the BFAP studies and pilot schemes. Many of the contributors to the BFAP seem to perceive the floods and their effects on the people in a very different way from those affected. Some Bangladeshi organizations and experts still consider that the approach is very top-down, and insensitive to what causes vulnerability (Adnan 1993; Farooque 1993). Because of the extraordinary coincidence of two very exceptional (perhaps 100-year) floods in 1987 and 1988, it seems an excuse has been found to try to end all floods, including those that most rural people consider to be beneficial and essential.

The plans fail to take adequate account of the fact that the flood prevention will produce its own set of victims, those who are going to be made worse off by the proposed projects. It is even possible that more people will suffer longer-term damage to their livelihoods as a result of the flood prevention projects than the number who suffer in flooding. It is essential to calculate the

Box 6.1 continued

impacts on the poor and vulnerable of any increased financial burdens the country may face as a result of the heavy borrowing likely to be required for the eventual construction projects.

The principle 'hard' projects for flood prevention suggested in the BFAP can be summarized as the construction of high embankments (or repair of existing ones) along much of the length of the main rivers (Ganges, Brahmaputra, Meghna: see Figure 6.2), combined with protection of land for High Yielding Varieties (HYV) Green Revolution agriculture. In addition there will be some new or improved anti-storm surge bunds constructed or repaired on low land at the mouths of the delta distributaries. These are designed to protect farm-land from salt-water incursion during cyclones (these are not in the worst-hit areas of the 1991 cyclone, on the east coast). The 'soft' components include the development and improvement of flood warning systems. (Summaries of the projects are to be found in Boyce (1990); Brammer (1990b); Dalal-Clayton (1990).)

The basic conception of the BFAP has been that water will be contained and moved downstream between large, long embankments, despite the fact that there is little knowledge of the downstream consequences (including possible flooding in other areas). The newly-protected HYV lands behind these banks are to be separated into 'compartments', surrounded by their own embank-ments. It is proposed that these compartments could be deliberately allowed to flood, in a version of the idea of living-with-floods. The idea is that water can be controlled to allow optimum levels for irrigation in the wet season, but with the ability to keep out flood peaks. Some such Flood Control Drainage and Irrigation (FCDI) projects have been in operation already for more than a decade, and have not solved poverty nor even removed all vulnerability to flooding.

The issue of land tenure and the distribution of assets and income generated by such schemes will not be dealt with in the projects. Even if such FCDI schemes work properly and increase agricultural output, the social factors which prevent poor people from getting adequate nutrition are not addressed. Increased output is not a necessary condition of solving hunger, and there is no guarantee that better production under flood prevention will help either, since the problem of hunger is not lack of food, but the inability of people to grow their own (land and asset poverty) or buy it (income poverty).

Supporters of the plan argue that the BFAP is not monolithic and is able to adapt to the most appropriate policies, in sympathy with the needs of the people. Critics argue that alternatives which might work better and be much cheaper are being ignored, and that the existing knowledge of many Bangladeshi and international NGOs is not being properly considered.

To understand the gap between these perspectives, we need to see how they fit into existing power systems. Disaster reduction is not isolated from other aspects of life in Bangladesh. It operates in a hierarchy which connects the vulnerable and poor in the village to national and international interests. It involves either ignoring or failing to understand the factors which generate people's vulnerability to floods. The existing distribution of power, income, and assets is a major component of that vulnerability. Numerous NGOs and people's organizations in Bangladesh are concerned that the proposed projects

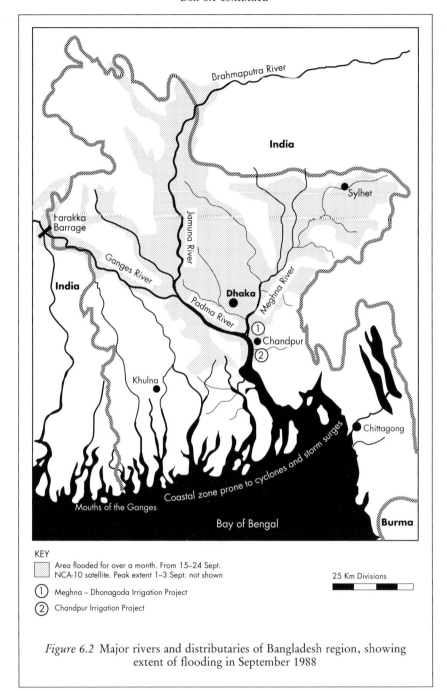

KEY

▨ Area flooded for over a month. From 15–24 Sept.
NCA-10 satellite. Peak extent 1–3 Sept. not shown

① Meghna – Dhonagoda Irrigation Project

② Chandpur Irrigation Project

25 Km Divisions

Figure 6.2 Major rivers and distributaries of Bangladesh region, showing extent of flooding in September 1988

(and even the pilot studies) reflect both a continuation of existing processes of generating vulnerability, and a failure to deal with the needs of the majority (Adnan 1993; Farooque 1993). One British member of the panel of experts advising on the plan concedes that consultation and public participation 'is not an easy task in a strongly hierarchical society, and it would be unrealistic to expect overnight success' (Brammer 1993: 9).

There are two interlinked factors involved in setting the BFAP on its course to a 'tech-fix' approach which are unlikely to change very significantly as future projects and investments are proposed. The first is the availability of the 'technical-fix' itself, as a set of expensive techniques involving major engineering works. It reinforces the benefits of the power system for those already in control, both in Bangladesh and internationally. Because it involves large-scale engineering contracts, foreign donors or lenders receive back a very substantial share of the spending, through consultancy fees and purchase of equipment. Likewise, politicians and others in the local elite benefit from kickbacks on the contracts, consultancies, and brokerage fees for the donor agencies in arranging local projects. Other ways of doing things are not nearly so attractive to either group, because they would involve less spending.

Secondly, preference for the 'tech-fix' approach is also linked to the desire of local elites to protect their own land and property, especially in towns and cities, and in areas of existing Green Revolution projects. One researcher recorded the following opinion:

> 'Some would argue', says Dalal-Clayton, that the flood plan amounts to 'a political response to the clamour for flood protection from wealthier, influential, urban-based groups . . . a great many people who presently live along the main rivers and on islands in the river channels will be exposed to increased risks from flooding'.
>
> (Quoted in F. Pearce 1991: 40)

The national priority is to raise agricultural production (without asset or income redistribution). Paradoxically, the worst floods in 1987 and 1988 have led to the highest harvests, as moisture remained in the soil in the dry season, and non-flooded areas received more rain. But the real problem for those who focus on raising output is that both low-rainfall monsoons in summer and the normally dry winters lead to low yields, except in areas with irrigation. It is extremely difficult to calculate, but it is conceivable that the losses from low water availability in winter, and from summers with poor rainfall, are greater than those which accompany floods. In a sense the need is more to deal with water shortages and *drought control* than flood prevention. Such projects tend to benefit middle peasants and wealthier large landowners. It is to be expected that new ones created under the flood protection works will benefit the same interest groups. Poorer farmers and the landless may benefit in this from higher demands for labour and a general rise in output, but this seems to be a secondary consideration in the prospective designs. Similar FCDI projects already exist, for example, around Chandpur south-east of Dhaka. Some sections of the poor, including those vulnerable to floods, benefit from the protection.

Box 6.1 continued

The issue is whether it is the best way to protect them, and whether the impact of protection and flood prevention on others will be less beneficial. In rural areas, many who are already vulnerable to flooding may well not be resited safely when embankments are built. The impact on transfer of water by the river-training measures is also unpredictable, and other areas are likely to be made more prone to flooding. In addition, it is likely that the asset and income position, and livelihood opportunities (especially for those who depend on common property resources), will deteriorate. There is widespread concern that flood protection works may eliminate the ponds that remain behind after floods in which fish and shrimps spawn. These 'common property' fisheries are of immense value to poor people both for nutrition, and as a source of income for those who catch and sell on to others. It has been estimated that inland fisheries account for 77 per cent of the national catch (figure quoted in Boyce 1990: 423).

The prevention of flooding may also seriously diminish the renewal of soil fertility that is brought about by the inundation, the resultant decomposition of plant residues, and an associated bloom of nitrogen-fixing algae (Brammer 1990b: 164). There is a further potential significant loss, of 'free' groundwater replenishment. In place of the natural recharging of the water-table by normal floods, the impact of the compartmentalized agriculture will be to require expensive pumping from rivers and channels to provide for both winter and summer agriculture.

Maintenance of the extensive and large constructions required by the BFAP will be both difficult and expensive. Commentators on the problems of the plan have pointed out that past experience of repairs to embankments – which is often poor – shows that the 'tech-fix' is unworkable. This is an important argument, because any failure of a major embankment will have appalling consequences, especially as river-beds are likely to rise above the level of the surrounding countryside as they become deposition zones for river silt. Even the impact on the compartmentalized FCDI agriculture could be serious if the banks fail. One similar FCDI scheme in existence at Meghna–Dhonagoda (to the north of the Chandpur one) suffered serious losses in 1987 because of a breached embankment (Thompson and Penning-Rowsell 1991: 6). Another likely maintenance problem, widely observed with existing flood prevention banks, is blocked drainage conduits preventing the return to rivers of impounded flood- and rain-water. Roads and railways are normally elevated on embankments to avoid inundation, but these structures commonly cause flooding through not allowing proper return drainage.

Alternatives to the plan: a people's needs-based approach?

Although the BFAP is intended to diagnose problems and seek appropriate solutions, there is widespread concern in the country and outside that it will automatically prescribe a 'tech-fix' approach. The main criticisms concern the costs involved (in relation to the uncertainties of the claimed benefits), the uncertainties of success, the increased dangers of failure (a breached embankment may cause a worse disaster than a bad 'natural' flood), the ignoring of

Box 6.1 continued

alternatives, and the failure to start dealing with the problem by first finding out how the majority of affected poor people themselves define it.

The emphasis of the USAID report (Rogers, Lydon, and Seckler 1989) is on managing the flood, or living with flood, rather than attempting expensive prevention. They are undoubtedly influenced by the many Bangladeshi scientists, social scientists, and NGOs who advocate modified 'living-with-flood' strategies. Boyce has suggested a number of alternatives to flood prevention, including the use of constructed ponds for the storage of water and fishing, better preparedness, and the building of low embankments that protect crops in early stages of growth while not preventing high floods (Boyce 1990: 421). But what is really crucial to understanding the problem is to ascertain what causes people to be vulnerable to flooding, and therefore to assess what measures can reduce vulnerability. These may be far removed from the need to prevent floods through measures, the costs of which the poor will have to bear disproportionately anyway. Figure 6.3 maps out some of these components of vulnerability, and it is there that the beginnings of river and rainfall flood disaster avoidance should be found. To protect people and their homes there needs to be investment in flood shelters (including elevated schools and health centres) and raised housing sites. To protect their health (and thereby indirectly their livelihoods) there needs to be protected water supplies for drinking, and primary health care (some authorities also advocate safe flood-proof latrines). To protect their livelihoods there needs to be access to land, replacements for land lost through erosion, compensation for losses of animals and other production assets. Where necessary, there must of course also be flood prevention measures, but based on an assessment of what is actually needed to reduce vulnerability rather than some grand design which supposedly prevents all flooding without considering who benefits and who loses, who pays for it, and whether it is needed anyway.

In 1991 tens of thousands died as the immediate consequence of flooding caused by a tropical cyclone on the southeast coast of Bangladesh (see Chapter 7). By contrast, the river and rainfall floods of 1987 and 1988 (in extent the worst experienced in memory) reportedly each killed less than 2,000 people directly (Brammer 1993; World Bank 1990: 40). One component of the BFAP deals with cyclone protection. But there seems to be much less concern for measures which can provide coastal villagers with effective cyclone shelters and other forms of protection. While the desire to reduce flood disasters is laudable, and tens of millions of people endured intense suffering in 1987 and 1988, few lives were lost, and the hardship remains in non-flood years: the basic vulnerability issues need to be addressed not through prevention of floods, but through changes in the processes that create the unsafe conditions. Starting there is likely to produce different approaches that may be less costly and more effective.

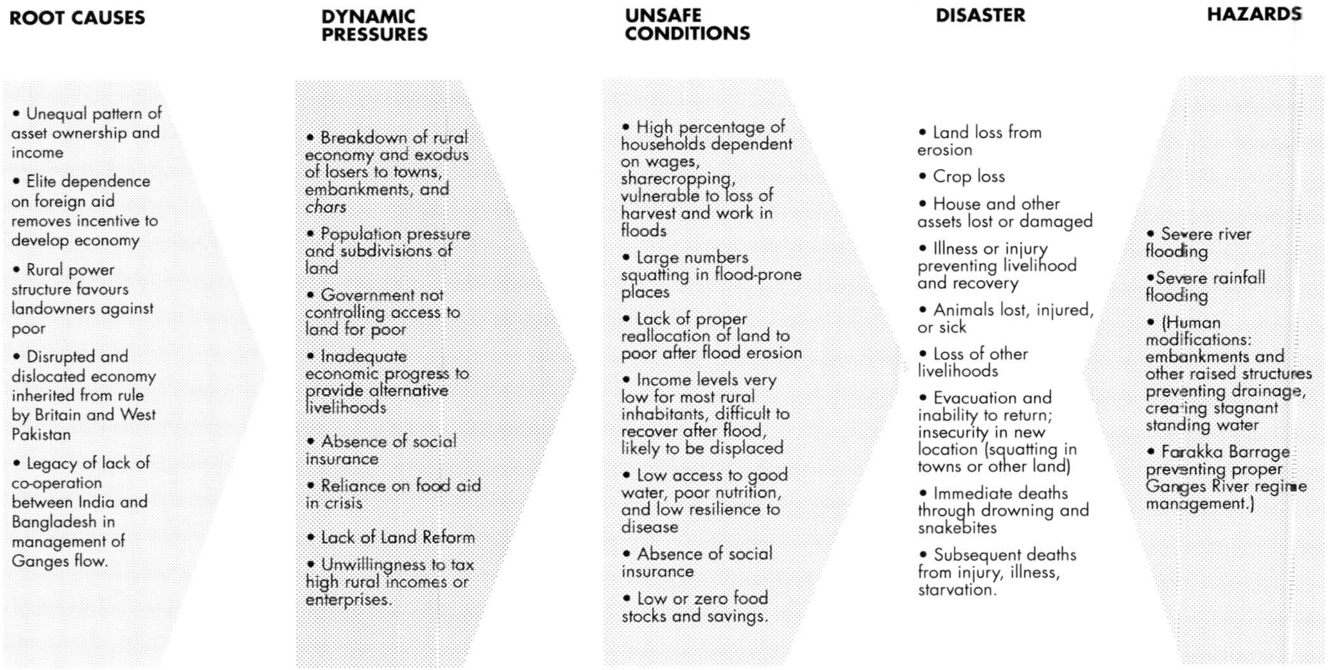

ROOT CAUSES	DYNAMIC PRESSURES	UNSAFE CONDITIONS	DISASTER	HAZARDS
• Unequal pattern of asset ownership and income • Elite dependence on foreign aid removes incentive to develop economy • Rural power structure favours landowners against poor • Disrupted and dislocated economy inherited from rule by Britain and West Pakistan • Legacy of lack of cooperation between India and Bangladesh in management of Ganges flow.	• Breakdown of rural economy and exodus of losers to towns, embankments, and chars • Population pressure and subdivisions of land • Government not controlling access to land for poor • Inadequate economic progress to provide alternative livelihoods • Absence of social insurance • Reliance on food aid in crisis • Lack of Land Reform • Unwillingness to tax high rural incomes or enterprises.	• High percentage of households dependent on wages, sharecropping, vulnerable to loss of harvest and work in floods • Large numbers squatting in flood-prone places • Lack of proper reallocation of land to poor after flood erosion • Income levels very low for most rural inhabitants, difficult to recover after flood, likely to be displaced • Low access to good water, poor nutrition, and low resilience to disease • Absence of social insurance • Low or zero food stocks and savings.	• Land loss from erosion • Crop loss • House and other assets lost or damaged • Illness or injury preventing livelihood and recovery • Animals lost, injured, or sick • Loss of other livelihoods • Evacuation and inability to return; insecurity in new location (squatting in towns or other land) • Immediate deaths through drowning and snakebites • Subsequent deaths from injury, illness, starvation.	• Severe river flooding • Severe rainfall flooding • (Human modifications: embankments and other raised structures preventing drainage, creating stagnant standing water • Farakka Barrage preventing proper Ganges River regime management.)

Figure 6.3 'Pressures' that result in disasters: Bangladesh floods 1987 and 1988

Flood avoidance measures

Where there are known river flood hazards, land-zoning measures can be effective in preventing disaster by literally avoiding the flood, or giving priority in land-use in flood-prone areas to certain types of users. Unfortunately, it is common in developing countries for there to be a large number of people who ignore or evade such restrictions. Squatting on unstable hillslopes which can collapse in heavy rain, or in low-lying flood-prone areas, is often the only way for the poor to obtain any land for housing (see Chapter 8 on landslides).

Flood mitigation and preparedness

Mitigation policies can save lives and protect property even though the flood itself cannot be prevented, contained, or avoided. The most conventional of such preparatory methods is flood warning systems, the effectiveness of which has been shown in a range of countries. The value of warnings depends greatly on their accuracy (this affects their credibility), the lead time available for preparedness and evacuation, and the effectiveness of the message delivery system.

The delivery and receipt of the warning messages cannot be taken for granted. There have been cases where, as with the Andhra Pradesh (east India) cyclone in 1977, warnings may be issued but not received by the most vulnerable sections of the rural population. They became victims because neither they nor their neighbours possessed a transistor radio on which to hear the relevant broadcasts.

NOTES

1 Patterns of vulnerability to floods may also be relevant for areas prone to tropical cyclones. The more detailed discussion of tropical cyclones will be left to Chapter 7, despite the fact that they are often a proximate cause of major flooding. We make this separation mainly because there are other distinct forms of damage involved with cyclones (especially wind), and also the precautionary measures needed to reduce vulnerability and risk are different.

2 The reservoirs created by dams eventually silt up so that the dam becomes ineffective for power generation, irrigation, or flood control, while the submerged land cannot be recovered. There is considerable controversy over the rates of siltation, and designs have often been more optimistic than actual performance. In many cases, a lifespan of forty years is normal. On the Narmada, see Alvares and Billorey (1988).

3 This mortality incorporates all types of flooding, including that from tropical cyclones: it is difficult with available data to separate different types of floods. However, although deaths associated with cyclones have on some occasions been very high (e.g. Bangladesh, possibly 224,000 in 1970), other floods are far more frequent and widespread. A few single events have been responsible for millions of deaths, for instance, an estimated 2.5 million with the Huang River flood in China, 1887.

4 These people from southern Sudan have endured disruption of their lives by war and drought, causing them to seek refuge (probably in long, difficult stages) near Khartoum; then the floods and other risks of closely packed, unsanitary living; and finally, forced expulsion into the desert! It is heartbreaking to imagine individuals or families actually living through such a series of disasters. The complex issues are discussed partially in Chapters 4, 5, and 9.

5 This is an unusually warm current in the Pacific off the coast of South America that brings torrential rain when it appears. For a discussion of El Nino see Chapter 7.

6 Field observations by Cannon in Haryana and Uttar Pradesh, north India, in 1976 and 1979.

7

SEVERE COASTAL STORMS

INTRODUCTION

People have lived along coasts since antiquity. The most recent phase of colonial expansion (since the middle of last century) and the establishment of a world market have greatly increased the numbers of urban settlements, plantations, ports and naval bases, and other centres of population in coastal areas. More recently, tourism and the global expansion of export-oriented industries have added to the attraction of coastal locations. Of the world's major cities in 1980, 66 per cent of those with over 10 million inhabitants, and 53 per cent of cities of those with between 5 and 10 million, were situated on the coast (United Nations 1980). Several of the fastest growing cities, all projected to have 20–30 million inhabitants by the year 2025, have long histories of exposure to severe tropical storms. These include Karachi (Pakistan), Jakarta (Indonesia), Calcutta (India), and Dhaka (Bangladesh) (Davis 1986: 279).

The 'attractiveness' of coastal locations is often 'taken for granted'. For instance, Griggs and Gilchrist review numerous high-risk situations on the Gulf and Atlantic coasts of the US and conclude that 'people want to live in the sun and be able to look at the ocean; realters and developers want to make money, and local governments want more tax dollars' (1983: 274). Burton, Kates, and White (1978: 4–17) emphasize the attractiveness of rich alluvial soils in the case of coastal Bangladesh, a factor that is significant in many other regions.

In this chapter we examine the reasons why people inhabit coasts and their implications for vulnerability in the face of hazards.[1] In seeking the causes of vulnerability to cyclones, typhoons and hurricanes (these being different regional names for the same climatic phenomenon) such generalizations need to be investigated more specifically. Cyclonic storms do not affect all coasts equally. Where such storms are frequent, not all people suffer equally. Where people suffer, not all people are able to reconstruct their lives rapidly or equally well. These more detailed questions about vulnerability are applicable to storms affecting rich countries like

Australia, Japan, and the US as well as poor countries like Fiji, Mozambique, Nicaragua, Bangladesh, and the Philippines. For every 'voluntary' resident in a high-risk coastal location (those seeking 'sun and surf', for instance) there are thousands who have no alternatives because their livelihoods are tied to jobs in oil refineries or export enclaves, to jobs in the service sector spun off by the tourist trade, to jobs on fishing boats, or to employment on coastal farms and plantations.

Patterns of death and damage due to these storms and the ability of people to reconstruct their livelihoods show differences according to the national wealth, history, and socio-political organization. Recovery following hurricanes in the Caribbean in 1988 and 1989 showed such contrasts. Nicaragua mobilized a nationwide effort to help victims of hurricane Joan on that country's Atlantic coast (Nicaragua Ecumenical Group 1988). In part, this was an opportunity for the Sandinista government to build support among sections of the people who had not benefited substantially from the 1979 revolution, and in some cases had supported the opposition Contras.

By contrast, relief efforts in Jamaica following the 1988 hurricane Gilbert were rife with partisan politics and corruption, so much so that mismanagement was one of the factors that led to a change in government in the elections that followed. In the US, the major obstacle to relief and reconstruction on the South Carolina coast following hurricane Hugo in 1989 was bureaucratic blindness to the needs of (often illiterate) poor people who lacked insurance and other support systems (Miller and Simile 1992). Many of these marginal people, who may never recover from hurricane Hugo, are African Americans. Two decades earlier, when hurricane Camille devastated the Mississippi delta, the US Senate investigated charges that relief assistance had been racially biased (Popkin 1990: 124). Similar problems have been identified by Laird (1992) among poor, Spanish-speaking people near the epicentre of the 1989 earthquake in northern California (see Chapter 8). The US experience of such storms suggests then that planners attempting to reduce vulnerability would need different strategies than those appropriate in Jamaica or Nicaragua.

Likewise, the extraordinary differences in mortality from similar physical events should alert planners, citizen activists, and development agencies to significant differences in preparedness, response, and vulnerability. Australia suffered two very similar cyclones shortly after the catastrophic 1970 storm in the Bay of Bengal that killed 300,000 in Bangladesh (Carter 1987: 490). Yet the death toll in Australia was less than 100 (Stark and Walker 1979; Western and Milne 1979).

Socio-political organization can be as significant as national wealth in disaster preparedness. For example, in 1971 North Vietnam survived a combination of a coastal storm surge and torrential rain in the Red River delta that could have cost as many lives as in Bangladesh in 1970. But only a few hundred lives were lost in North Vietnam, largely because of highly efficient wartime village-level organization that allowed rapid evacuation and provision of first aid (Wisner 1978a).

Again, in 1974, when cyclonic storms delivered equivalent amounts of rain and wind in two parts of the world, forty-nine were killed in Darwin, Australia (Western and Milne 1979: 488) and 8,000 in Honduras (CIIR 1975: 1). The major difference seems to have been the pattern of rural landownership in Honduras, where 63 per cent of the farmers had access to only 6 per cent of the arable land (CIIR 1975: 13). Large-scale beef ranches and banana plantations had displaced peasants over several decades into isolated valleys and steep hillside farms. Here they received little warning and were at risk from mudslides that accompanied the hurricane's rainfall. Deforestation by these peasants seeking to carve out subsistence farms had made the hillsides unstable. In the northern town of Choloma, 2,300 people were killed when an artificial dam created by landslides into a nearby river burst, sending masses of black mud into its streets (CIIR 1975: 3).

THE PHYSICAL HAZARD

Tropical cyclones are one of the most powerful atmospheric phenomena. A fully developed hurricane releases the energy equivalent of many Hiroshima-sized atom bombs (Cuny 1983; Milne 1986: 71). These storms arise during the summer over various oceans in a belt north and south of the Equator. In addition to the wind damage and flooding caused by cyclones, there is a wide variety of possible physical effects involving a web of social and natural linkages. Wind and wave action have immediate impacts, but erosion and salt-water incursion can handicap the economy for months or even years (J.R. Campbell 1984). Damage to roads, telecommunications, and power facilities can have short- and long-term effects, and complicate other problems. Even in areas remote from the coast, associated heavy rainfall can provoke mudslides and other mass movements. In the next section we will discuss the differential vulnerability of various social groups to the various kinds of physical damage.

Tropical storms are seasonal yet highly unpredictable. Year-to-year severity and frequency of storms may be related to factors working at the global atmospheric level, such as ocean current changes. There is additional uncertainty because the direction, speed, and growth dynamics of such storms have not yet been understood despite heroic attempts at computer modelling. As a result, warnings broadcast over the media sometimes result in needless evacuations, making it more difficult to convince the public on later occasions. For instance, in the aftermath of the 1970 cyclone in Bangladesh (then East Pakistan), many warnings – some of them false – were issued over the radio. Some authors claim that the losses due to the 1985 cyclone might have been lower if the public had been less cynical about broadcasts of warnings

(Milne 1986: 73–4). On the other hand, millions of people did respond to warnings of the great cyclone that hit Bangladesh in 1991.

PATTERNS OF VULNERABILITY

Coastal locations were frequently the first point of contact between indigenous people and colonial powers. These footholds, established from the sixteenth to the nineteenth centuries (first in Latin America and Asia, later in Africa), often grew into major urban centres. Livelihoods all along the new corridors of administration and extraction became dependent on needs of the colonizer (e.g. the market for labour power, groundnuts, cotton, beef, etc.) (Franke and Chasin 1980). Squatter urbanization began around the newly-established coastal cities (Hardoy and Satterthwaite 1989).

Migration to cities has become commonplace in the post-colonial world. The percentage living in cities larger than 100,000 (many of them coastal) in 'less developed countries' increased from 3 to 20 per cent between 1920 and 1980 and is projected to reach 30 per cent by 2000. In the 'more developed countries' some 21 per cent already lived in such cities in 1920, and by 1980 this percentage had reached 47 per cent. That number may reach 55 per cent by 2000 (Armstrong and McGee 1985). There is some regional variation, but the rapid growth of Third World urbanization is consistent (See Table 7.1). Urban populations remain in crowded areas, with many residing in low-lying, flood-prone areas, in flimsy housing, and with a lack of infrastructure. The millions who have caused these former colonial cities to swell into today's mega-cities are part of the patterns of extractive, export-oriented economic activity established a century or more ago.

Persistent urban poverty underlies the pattern of people's vulnerability. Hardoy and Satterthwaite (1989: 159) cite examples of urban settlement on hillsides prone to landslides in Rio de Janeiro (Brazil), Guatemala City, La Paz (Bolivia), and Caracas (Venezuela) and land prone to flooding or

Table 7.1 Population of cities over 100,000 in Third World regions

Continent	1920		1980		Projected 2000	
	Millions	*Per cent of total*	*Millions*	*Per cent of total*	*Millions*	*Per cent of total*
Africa	3	2	86	17	249	31
Latin America	9	10	158	43	355	57
East Asia	21	4	230	21	431	32
South Asia	13	3	204	14	539	24
Oceania	3	21	12	53	18	56

Source: Armstrong and McGee (1985: 7).

tidal inundation in Guayaquil (Ecuador), Recife (Brazil), Monrovia (Liberia), Lagos and Port Harcourt (Nigeria), Port Moresby (Papua New Guinea), Delhi (India), Bangkok (Thailand), and Jakarta (Indonesia). These urban poor have declining connections with rural relatives with each passing generation, so the resources available to them for reconstruction of livelihoods after extreme storms are also deteriorating.

Contemporary coastal settlement

In both wealthy countries such as Australia and the US and the poorest former colonies such as Bangladesh, Mozambique, and Jamaica, there are large towns and cities on hurricane-prone coasts. Potential economic losses are a function of this pattern of urbanization.

Cyclone Tracy (1974) destroyed 50–60 per cent of the houses in Darwin (Australia) and caused A$3.2 billion of damage to the city (Wilkie and Neal 1979: 473). If it had struck the coast just 100 km (60 miles) either side of that city, losses would have been hardly noticed (Stark and Walker 1979: 191). In 1969 hurricane Camille missed the major US city of New Orleans by about 100 km (60 miles). Even so, 262 people died and losses totalled US$1.4 billion (Petak and Atkisson 1982: 332). Hurricane Gilbert (1988) was the most powerful storm ever recorded in the Western Hemisphere (with a low pressure of 888 mb). It travelled directly over Kingston, capital of Jamaica, and traversed the full length of that island. The death toll was forty-five and damage was estimated at US$1 billion (Barker and Miller 1990).[2]

Rural areas in the tropics may contain capital-intensive plantation and other agro-industrial facilities. Jamaica lost 30 per cent of its sugar hectarage, 54 per cent of its coffee, and more than 90 per cent of its bananas and cocoa. Hurricane Gilbert thus deprived Jamaica of more than US$27 million in foreign exports in 1988–9 alone (Barker and Miller 1990: 111). Mozambique lost two cashew-processing factories during the cyclone that struck in 1979, in addition to many thousands of economically important trees (coconut and cashew) (Wisner 1979).

In countries with high rural population densities and considerable inequalities in income and access to land, rural concentrations of people in the high-risk coastal zones can be very great (for example, the Philippines, parts of Indonesia, the Sundarbans and islands such as Sandwip in Bangladesh, and India). In these cases, as we show in detail in the next two sections, local or even distant pressure on land forces the poor to remove protective vegetation, destroying buffer zones and increasing their vulnerability to storms.

But there are also vulnerable groups in richer countries like the US. A significant proportion of the newcomers to the coast of Florida are older retired people (Graff and Wiseman 1978). During hurricane Agnes in 1972 it was realized that they required special assistance during evacuations

(Briggs 1973: 134). Retirement homes, tourist and recreational centres have grown rapidly near and even quite literally on these coasts. An extreme Camille-sized hurricane hitting the major Florida recreation and retirement centre of Dade Country (near Miami) would destroy an estimated 18 per cent of the structures directly on the coast and up to 7 per cent as far away as 160 km (100 miles) (Cochrane 1975: 19).

Skilled workers in the US are sold the American dream of retirement in the sun, in cheaply-built properties in high-risk locations. This is another facet of vulnerability created by corporate production, sales, and investment strategies. It echoes the activities of US corporations which increase land pressure in countries like the Philippines through acquisition of land for growing pineapples, bananas, or other export commodities. The small farmers thus dispossessed are forced to live in dangerous urban coastal locations through the actions (directly or indirectly) of a transnational corporation (Boyce 1992). The skilled working class and middle class back in the US are tempted into similar hazardous zones as the value extracted abroad is cycled through the financial circuits of the system in the form of coastal real estate development.

Small island nations finding themselves in the path of frequent tropical storms should be considered separately. Their population growth rates have often been lower than in other exposed areas. Yet the ravages of a tropical storm can encompass the entire island and its people (J. Lewis 1981, 1984a, 1984b). For instance, hurricane Hugo (1989) destroyed 85 per cent of the housing stock on the Caribbean island of St Croix, and in 1972 hurricane Bebe left 20 per cent of Fiji's population homeless. The economy of such islands is often based on a small number of tree crops, such as bananas and nutmeg in many Caribbean islands and coconuts in Oceania (Shakow and O'Keefe 1981). The small-scale fishing sector is often also seriously damaged, and though generally more easily replaced than a coconut plantation, a poor fisherman may find it very difficult to replace a lost boat or equipment. Even with access to credit, a vicious spiral of indebtedness can result.[3]

Coastal livelihoods

People's livelihoods have already been mentioned in the course of the preceding demographic review. In most rural coastal areas, the political economy of access to resources is quite complex and fragmented. Even where large landowners, companies, or the state have rights to coastal plantations or fisheries, various kinds of formal and informal, legal and illegal access may be available to workers, the seasonally employed, and neighbouring poor people. Small-scale fishing tends to be controlled by owners of boats who employ a large percentage of fishermen, as in the coastal waters of northeastern Brazil, Sri Lanka, and Andhra Pradesh

(India). The trend towards concentration of ownership can also be seen in the affected US coastal fisheries, where (as with the 'family farm' in the US) high interest rates and changing production technology have put pressure on small, self-employed units. The latter tend to under-insure their equipment due to cash-flow pressures and are highly vulnerable to the economic effects of production interrupted by a hurricane.

Coastal land and marine resources worldwide tend to be under the increasing control of absentee interests. This is true of both developing and industrial countries. An economic calculus applied from a distance often leads to land-use decisions that put people and the sustainability of local ecosystems at risk. This is the case where remote financial interests continue to drive speculative residential development in Florida. The effect of similarly 'distanced' economic rationality can be seen in Haiti and coastal Kenya, where urban commercial interests encourage conversion of the last of the remaining mangrove wetlands to charcoal for sale as urban fuel (World Resources Institute 1986: 146–9), or in Jamaica, where it was recently proposed to mine large portions of coastal wetlands for peat as a substitute for imported energy (Maltby 1986: 135–44). Such actions remove vegetative barriers that anchor soil and absorb some of the force of storm winds and waves. In the Jamaican case it has been predicted that illegal cannabis production would be displaced from the Nigril Morass (wetland) to mountains further to the east. Deforestation of those mountains by cannabis producers would increase storm runoff and erosion (Maltby 1985) and could put the growers at risk from mudslides during severe storms such as hurricane Gilbert in 1988.

It is useful to distinguish between patterns of livelihood in the immediate vicinity of large coastal cities and those in more remote coastal areas. In the former case, vulnerable rural people may have relatives and livelihood options open to them in the city as a refuge when a storm hits. But the disruptive influence of the city over the immediate rural coastal area (e.g. loss of access to resources because of tourism, industrial pollution, or commercial market gardening) has to be weighed against the advantages of such a possible urban 'fall-back' position for rural relatives and kin.

The petty commodity producer and semi-subsistence farmer in many tropical coastal areas utilises a variety of complementary livelihood options. Examples include 'community forest' farming in Sri Lanka and the combination of fishing and farming in Fiji and other parts of the world. However, these multiple options have been undermined by government and commercial encouragement of monocropping, the growing importance of wage employment, and competition for resources from absentee commercial interests. Thus, in coastal India, Thailand, and the Philippines small-scale fishing for domestic consumption is declining due to competition with and overfishing by commercial trawlers in the same coastal waters (G. Kent 1987). In some areas it has been noted that the diversity of locally-adapted

153

food crops, including those most hurricane-resistant, has decreased (J.R. Campbell 1984; Pacific Islands Development Program n.d.).

The livelihood systems of the rural poor in exposed coastal areas are heavily influenced by spatial and temporal constraints. There is not only a connection between seasonality and rural poverty (Chambers, Longhurst, and Pacey 1981; Chambers 1983), but vulnerability has its own temporal rhythms (as noted in previous chapters). The timing of certain farming operations is critical to yields, and the plants and animals themselves are more vulnerable to damage due to flooding or wind at some times than at others.

More importantly, perhaps, wage work is most available at certain seasons. Typically, the harvest season sees large numbers of the rural poor engaged in temporary work on the harvest in fields and plantations. Tropical storms striking at this time of the year may mean large economic losses to the owners, and drastic reduction in income to workers who lose employment. There is a parallel with drought; in many cases it not only denies the small farmer or herder their subsistence, but it also deprives them and the rural landless of employment on larger farms (see Chapter 4). Riverine flooding in Bangladesh denies the landless and land poor employment in the rice and jute fields of the more prosperous (Currey 1984; Clay 1985; and discussion in Chapter 6).

Fishing is also seasonal, sometimes leading to complex risk calculations. If fish yields are best at greater distances from shore during the hurricane season, where should the 'rational' fisherman place his nets? The fisherman has a greater chance of saving the nets within the usual storm warning time if he sets them closer to shore, but will catch more by gambling on the more distant fishery (Davenport 1960; Gould 1969).

We can summarize the relationship of vulnerability, livelihood, and cyclones by using the access model in relation to hurricanes (see Figure 4.1). Beginning with the household and their stock of resources and assets (box 2), there are severe but differential impacts. Family members die and are injured, but more so among the poorer social groups. According to one account of the Andhra Pradesh cyclone, 75 per cent of the 8,000 deaths were in small farming, fishing, or marginal farming families (Winchester 1986). In Bangladesh in 1991 the wealthy survived in concrete-block houses (Khan 1991).

Loss of labour power in poorer households makes it harder for them to recover economically. Assets are also lost. The wealthy lose more in absolute terms, but less in relation to their total property. In particular, farm buildings, livestock, and perennial (tree) crops are lost as well as the standing annual crop. Salt-water flooding of fields may further complicate recovery (J.R. Campbell 1984). All of this modifies the household's income opportunities (box 3). They may be able to add a quick 'catch crop' if seed and other inputs are available. Otherwise tree crops may be missing from their list of

154

options for some years. Income from wage-labour in reconstruction may be available as a new option for some households.

Class relations and structures of domination (boxes 1 and 4) influence the new array of livelihood options. Poorer households will not be able to pay off loans or meet rent and other obligations. Credit provided by government and other sources may not be so easily obtained by the poorer social groups. Indebtedness and dependency will tend to reinforce the allocation of social power, hence the structures of domination. In this the access model differs somewhat from that applicable to a disaster such as famine. In that case the legitimacy of structures of domination has sometimes been called into question by the majority of the population.

The household budget (box 7) is radically changed. Indeed, for a period during relief and recovery the poorest households may simply not have a budget if they have lost everything – house, livestock, tools – and will be largely dependent on outside aid for food (as in Bangladesh following cyclones in 1970, 1985, and 1991). If aid is adequate, malnutrition and disease should not follow (boxes 10 and 11), but there is a chance that a weakened population can fall prey to epidemic disease due to poor sanitary conditions, especially the lack of safe drinking-water (see Chapters 5 and 6).

Large cyclones are not uncommon on this part of the Indian coastline. In 1864 an enormous storm surge killed an estimated 30–35,000 people. More recent cyclones occurred in 1927 and 1949, with two in 1969 (Cohen and Raghavulu 1979: 9). What made the 1976 situation unusual was the coincidence of the cyclone with high tide. This was compounded by the fact that the storm suddenly changed course, confusing meteorological forecasting so that there was no warning.

The worst-hit population lived in Divi Taluk, which lies at the mouth of the Krishna River between its two branches and the Bay of Bengal. Winchester (1986) mapped the deaths and found a high correlation between percentage deaths and the height of the storm surge (Figure 7.1). But his detailed analysis of topography showed that even within this relatively flat, alluvial delta, there were higher sites that were more secure from the storm surge (3 to 7 metres (10 to 23 feet) in height). For the most part, wealthy farmers and petty officials who lived on these sites in 'proper' (*pucca*) concrete houses survived. However, the whole community lacked public shelters of this kind, and those buildings perceived as safe (schools, temples, administrative headquarters) were located in low-lying sites. In addition to the immediate impact, poorer social groups were in a weak position to rebuild their livelihood systems, and their recovery was very difficult. They lost tools, draft animals, milch cows, and labour power.

Why would people expose themselves to such risks? Their situation is illustrated by the 'pressure' model for the cyclone (Figure 7.2). The pressures leading to vulnerability included land hunger and the desire by low caste farm labourers and fishing folk to live near their employment. Roads

155

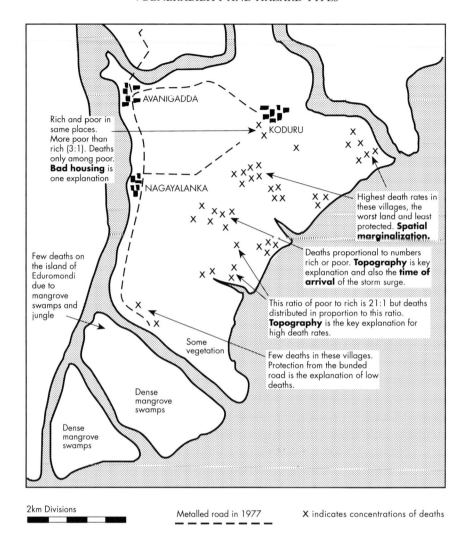

Rich and poor in same places. More poor than rich (3:1). Deaths only among poor. **Bad housing** is one explanation

AVANIGADDA

KODURU

Highest death rates in these villages, the worst land and least protected. **Spatial marginalization.**

NAGAYALANKA

Few deaths on the island of Eduromondi due to mangrove swamps and jungle

Deaths proportional to numbers rich or poor. **Topography** is key explanation and also the **time of arrival** of the storm surge.

This ratio of poor to rich is 21:1 but deaths distributed in proportion to this ratio. **Topography** is the key explanation for high death rates.

Few deaths in these villages. Protection from the bunded road is the explanation of low deaths.

Some vegetation

Dense mangrove swamps

Dense mangrove swamps

2km Divisions

Metalled road in 1977

X indicates concentrations of deaths

Figure 7.1 Explanation for deaths caused in the Andhra Pradesh cyclone of 1977
Source: After Winchester 1986.

and transport in and out of this administrative area (*taluk*) were poor, so labourers could not easily commute to work from higher land further inland. Finally, the ultimate causes of these pressures included the caste system, land tenure, and the nature of government. These three deeply embedded systems combined to allocate most resources to the wealthy farmers and petty officials. They also denied the area adequate investment in public shelters, all-weather transport to inland areas, and rural development assistance to the poorer social groups that might have enabled them to find livelihood options that did not demand their residence in exposed sites.[4]

Some lessons do seem to have been learned since then. Reddy (1991) reports that in May 1990 there was a remarkable example of successful evacuation of more than 500 villages as a result of a better warning system and improved government practical arrangements. It involved over 2,000 evacuation teams and more than 1,000 temporary relief camps. In all, 650,000 people were evacuated from the path of the cyclone, and the number of casualties was much less than in the same area in 1976.

COUNTRY CASE-STUDIES

Remote coasts

Mozambique has a coastline of more than 2,000 km (1,240 miles) along the Indian Ocean, all of it subject to tropical storms including some of cyclone force.

The colonial pattern of urbanization was highly centralized, and Lourenqo Marques (present-day Maputo), the capital on the coast in the extreme south, dominated all other centres. Thus, very long stretches of coast and the surrounding hinterlands were, and remain today, relatively remote from the three cities, except for small administrative towns with very few economic or social functions. These are connected by very poor roads, and communications are made even worse by the risk of bandit attacks that have increased since 1981.

Livelihoods on these stretches of more remote coast revolve around small-scale fishing, subsistence farming, limited livestock herding, remittance of wages from South Africa, income from limited sale of cashews, cotton, groundnuts, and food crops, and casual employment in plantations. The most important agro-industries in the coastal zone are coconut, cashew processing, and sugar cane.

After independence, some state farms were devoted to basic food crops (maize, rice, potatoes), and sunflower and cotton have also been developed. Wages in both state and private plantations are low, and much lower than migrant wages in South Africa. However, since independence South Africa has continually cut back on the numbers of Mozambican men it allows to work. Furthermore, producer prices for crops sold to the government have been kept low and, more importantly, commodities for purchase in the countryside have become increasingly scarce and expensive since the early 1980s. As a result of all these factors, much activity by the people of these more remote zones is focused on their own needs. Although there is limited local trade and barter, there is none on the national market.

In 1979 the central and northern coasts were battered by a cyclone that

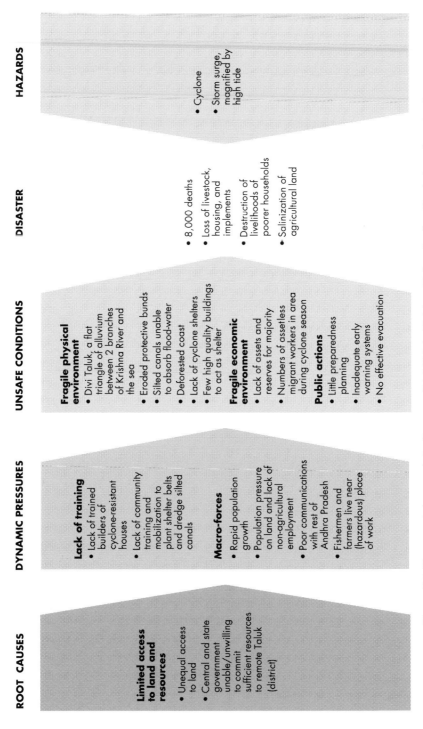

ROOT CAUSES

Limited access to land and resources
- Unequal access to land
- Central and state government unable/unwilling to commit sufficient resources to remote Taluk (district)

DYNAMIC PRESSURES

Lack of training
- Lack of trained builders of cyclone-resistant houses
- Lack of community training and mobilization to plant shelter belts and dredge silted canals

Macro-forces
- Rapid population growth
- Population pressure on land and lack of non-agricultural employment
- Poor communications with rest of Andhra Pradesh
- Fishermen and farmers live near (hazardous) place of work

UNSAFE CONDITIONS

Fragile physical environment
- Divi Taluk, a flat triangle of alluvium between 2 branches of Krishna River and the sea
- Eroded protective bunds
- Silted canals unable to absorb flood-water
- Deforested coast
- Lack of cyclone shelters
- Few high quality buildings to act as shelter

Fragile economic environment
- Lack of assets and reserves for majority
- Numbers of assetless migrant workers in area during cyclone season

Public actions
- Little preparedness planning
- Inadequate early warning systems
- No effective evacuation

DISASTER
- 8,000 deaths
- Loss of livestock, housing, and implements
- Destruction of livelihoods of poorer households
- Salinization of agricultural land

HAZARDS
- Cyclone
- Storm surge, magnified by high tide

Figure 7.2 'Pressures' that result in disasters: Divi Taluk, Krishna Delta, Andhra Pradesh cyclone 1977

destroyed two of three cashew-processing factories and uprooted 50,000 cashew trees (Wisner 1979: 304). At the time cashews were a major agricultural export. Apart from the cost of repairing these processing facilities, there were US$5 million in losses to the cotton and maize crops and to the system of feeder roads. Loss of life, however, was negligible.

The response to floods in 1976, 1977, and 1978 (the first three years of independence) had been to accelerate the government's programme of planned, voluntary resettlement in communal villages. The policy concentrated on people living in flood-prone locations (Wisner 1979: 202; also discussed in Chapter 6). As many as 500,000 people had been successfully resettled in more secure locations, many of whom were safe in 1979 as a result.

Earlier in this chapter we noted that social mobilization of this kind had been effective in saving lives in Vietnam (Wisner 1978a). Maskrey (1989: 79–86) reviews what he calls community-based mitigation programmes. He discusses other positive cases, including the response to floods in western Nicaragua in 1982 by neighbourhood 'committees for Sandinista defence' (citing Bommer 1985). We will return to the issue of community-based mitigation in Part III of this book. Here we provide a summary model of 'release' (Figure 7.3) from the disaster that results when a hazard combines with vulnerability: it is the counterpart of the 'pressure' model applied to the Andhra Pradesh cyclone above.

In 1979, the government's response to the cyclone showed the optimism and energy of a newly-independent nation with swiftly-organized government relief. The president and cabinet ministers toured the affected areas within days, and a nationwide collection of donations reached into factories and schools, with 'nation-building' as the positive message to overcome the tragedy. The cashew industry was rebuilt, peasants were rehoused in a massive effort to accelerate the programme of communal villages in the affected areas, public health and nutritional standards were maintained, and livelihoods were rebuilt.

In 1984, by contrast, cyclone Daimone slammed into the Mozambican coast at a time when the people and government had already been trying to cope with drought for two years, while also dealing with increasing sabotage and massacres by South African-backed RENAMO mercenaries. The storm itself was similar to the one in 1979, and the cashew and coconut industries suffered again. It also flooded the largest sugar plantation in the country, Sena Sugar Estates, whose system of protective dykes had fallen into disrepair.

The government's difficulty in responding in 1984 compared with the success of 1979 is explained by a number of factors. Firstly, in 1979 there was already considerable foreign assistance on the ground because of the drought that had affected some parts of the country for four years. Feeding camps had also begun to respond to the needs of the internally displaced

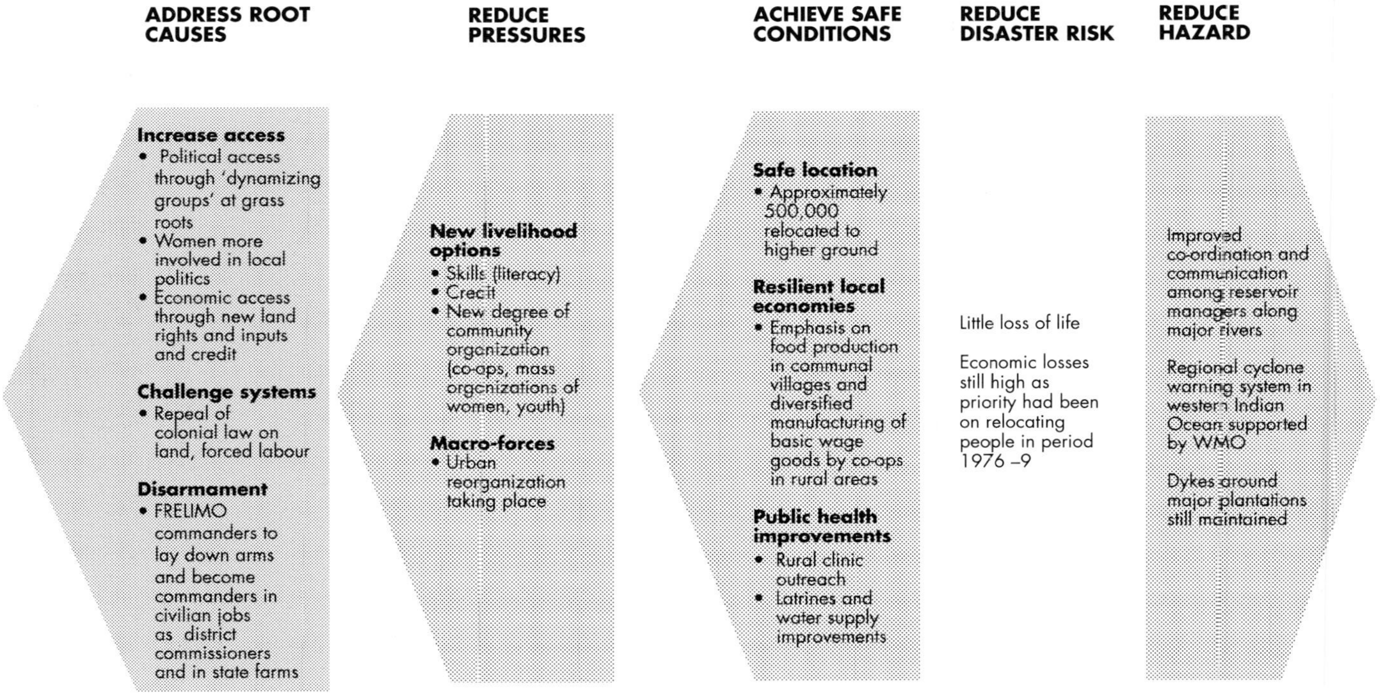

ADDRESS ROOT CAUSES

Increase access
- Political access through 'dynamizing groups' at grass roots
- Women more involved in local politics
- Economic access through new land rights and inputs and credit

Challenge systems
- Repeal of colonial law on land, forced labour

Disarmament
- FRELIMO commanders to lay down arms and become commanders in civilian jobs as district commissioners and in state farms

REDUCE PRESSURES

New livelihood options
- Skills (literacy)
- Credit
- New degree of community organization (co-ops, mass organizations of women, youth)

Macro-forces
- Urban reorganization taking place

ACHIEVE SAFE CONDITIONS

Safe location
- Approximately 500,000 relocated to higher ground

Resilient local economies
- Emphasis on food production in communal villages and diversified manufacturing of basic wage goods by co-ops in rural areas

Public health improvements
- Rural clinic outreach
- Latrines and water supply improvements

REDUCE DISASTER RISK

Little loss of life

Economic losses still high as priority had been on relocating people in period 1976–9

REDUCE HAZARD

Improved co-ordination and communication among reservoir managers along major rivers

Regional cyclone warning system in western Indian Ocean supported by WMO

Dykes around major plantations still maintained

Figure 7.3 The release of 'pressures' to reduce disasters: the Mozambique cyclone 1979

population seeking refuge from the increasing number of attacks by RENAMO.[5] This existing infrastructure was able to help storm victims immediately. By 1983–4 RENAMO attacks were responsible for numerous communication breakdowns, making response to the cyclone more difficult. Also economic disruption by RENAMO in the 1980s had been so great the government had fewer resources with which to respond.

Nicaragua faced a similar challenge in 1988, when hurricane Joan virtually destroyed Bluefields, the largest city on the Atlantic. Eighty per cent of the buildings and infrastructure were destroyed as well as 95 per cent of all structures on Corn Island. 'Joan' flooded land for hundreds of kilometres along Nicaragua's river border with Costa Rica. This affected important beef-production areas, and caused flooding and landslides due to torrential rains in the rich coffee-producing central mountains. Losses were estimated to have been more than those of the 1972 earthquake (Nicaragua Ecumenical Group 1988).

Until the revolution in 1979, the Atlantic coast of Nicaragua had never been developed as part of the country. The population is a mixture of indigenous people and migrants from the Caribbean islands. English is more common than Spanish, population density is low, and livelihoods centre around fishing and extraction of products from the rain forest that covers much of the region (Ballard 1984).

The hurricane came at a time when Nicaragua's Sandinista government had been forced to fight a war against the 'Contras', who were armed and encouraged by the US. But unlike Mozambique, the war was shifting in favour of the government, and it was possible to mobilize the nation and provide considerable aid for the stricken region. Ironically, it was a fall in the world market price of coffee (by about 30 per cent) in 1989 that ultimately did more harm to the national economy. The economy was already weakened by the economic embargo imposed by the US and by years of war against the Contras. The dual economic shock of hurricane Joan and the crash of coffee prices was important in turning voters against the Sandinistas before the eventual change in government. Remembering the 'global pressures' introduced in Chapter 2, we can reflect that even though production losses due to the hurricane were very large, the politics and economics of US opposition to the Nicaraguan government did more harm to the economy.

Densely populated rural coasts

The cyclone that hit the Indian coastal state of Andhra Pradesh in 1977 exemplifies the interaction of physical hazard and human vulnerability (Winchester 1986, 1992). The storm surge killed between 8,000 and 12,000 people (accounts vary), with a gigantic storm surge that raised a mass of water 3–6 metres (9–20 feet) high, 80 km (50 miles) long, and 24 km

(15 miles) wide (Cohen and Raghavulu 1979: 1). Who were the people killed? Winchester (1986: 184) shows death rates of between 23 and 27 per cent for small farmers, fishermen, and marginal farmers and only between 3 and 4 per cent for large farmers and petty officials. The ability to evacuate by road, type of house and its strength of construction, and small variations in topography within villages were reasons why the wealthier farmers and petty officials survived to a much greater degree.

In fact the wealthier lost more property (housing, stalls, cattle, carts, and standing crops) than the poor, but lost a smaller proportion of their total property. Their housing was more strongly constructed. They and the petty officials were able to evacuate people and valuables in motorized vehicles to which they had access.

Another major aspect of differential vulnerability was the capacity of different households to recover from repeated shocks. Winchester (1986: 228) investigated rates of economic recovery from the 1977 cyclone among six occupational groups up to 1981. After four years large farmers lost most, but recovered more quickly and ended up better off. Small farmers recovered more slowly and were also somewhat better off. Petty officials were also more prosperous in 1981. By contrast, marginal farmers, fisher-people, and landless labourers barely regained pre-cyclone levels of prosperity.

Physical reconstruction of shelter cut deeply into the limited resources of all but the wealthiest groups. Winchester (1986) criticizes credit facilities, suggesting that government plans for building more secure housing were likely to make matters worse by emphasizing concrete-block construction that the poor could not afford, rather than lower cost improvements that would be feasible (a point also emphasized in the literature on post-earthquake reconstruction; see Chapter 8).

Bangladesh is also unfortunately very familiar with severe cyclones. The account of the storm that developed in the Bay of Bengal and struck the coast of Bangladesh in 1970 is well-known. The storm surge coincided with high tides which came at night. Many thousands of poor migrants were sleeping in the fields they had come to harvest in coastal areas. Most at risk were those in fields on low-lying silt islands (*chars*) in the many mouths of the gigantic delta of the Ganges and Brahmaputra Rivers (see Figure 6.1). Mortality was estimated at between 600,000 and 700,000 by Burton, Kates, and White (1978) and 300,000 by Carter (1987).

Bangladesh had experienced numerous cyclones, suffering great losses to severe storms in 1961, 1963, and 1965. The 1963 cyclone swept the coast from Cox's Bazaar to Feni, killing 80,000; and the death toll in 1965 was 18,000 (Hanson 1967: 12–18). Recent cyclones include that in 1985 which killed 10,000 and destroyed 17,000 homes, damaged another 122,000, and swept away some 140,000 cattle and 203,000 ha (500,000 acres) of rice and jute (Milne 1986: 74). Another large cyclone hit Bangladesh in December

1988, only months after rivers had overflowed to flood the largest area ever recorded, displacing millions of people (see Chapter 6).

On 30 April 1991 a cyclone larger than the one in 1970 crossed the coast near Chittagong, and killed tens of thousands of people. Residents of islands were most vulnerable, along with those living on the coast from Cox's Bazaar north to opposite Sandwip Island. At least half of the 10,000 people on Sandwip Island were killed.

While the magnitude of the human loss and destruction is hard to imagine, it is considerably less than the toll in 1970. Since then thousands of Red Crescent volunteers with radios have been trained to listen for and to spread evacuation warnings. The government had spent US$3 million on improving its warning system since 1985 (Mahmud 1988), and many hundreds of thousands of people did make it to safety. Likewise, public storm shelters built since 1970 fared well, though the problem was that there were only 302 of them. Ten thousand such shelters (costing US$5,000 each) are needed to protect the exposed population. Public works programmes have also been systematically building embankments, but these are generally too low to protect against the greatest cyclones. Raising embankments to the required height of 6–8 metres (18–25 feet) costs US$25,000 for each 30 metre (100 feet) length (Kristof 1991).

Such efforts are aimed mainly at achieving improved safety for those at risk in the face of the hazard itself, and to some extent at modifying the intensity of the wind in specific places. But although welcome and supported or even initiated at the grass roots, they are inadequate and puny compared with the numbers of people who need protection. As Figure 7.4 illustrates, nothing is being done further back along the chain that creates these unsafe conditions. The pressures and root causes (which are very similar to those shown in Figure 6.3) are not being addressed. Without the political will and significant changes in the national and international factors that affect Bangladesh, the efforts at local level are likely to remain inadequate (see also Box 6.1).

Studies of climate change predict that an even larger proportion of Bangladesh's population would be vulnerable to severe storms if global warming fuels more violent storms and causes a rise in sea-level that reduces the country's area. In the face of this grim future, there is considerable grass-roots effort to reduce vulnerability. We return to these endeavours and the vexing question of population pressure in the final part of the book.

The small island

Fiji lost about 30 per cent of its agricultural production capacity in 1985 when four hurricanes swept these 361 islands in the space of two months. Tree crops suffered losses of up to 80 per cent, and sugar-cane production

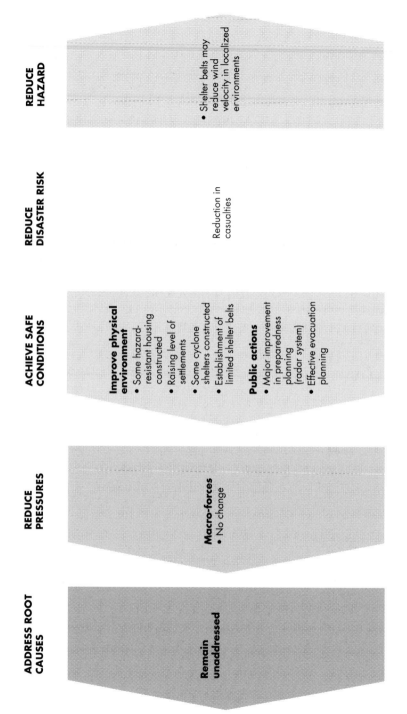

Figure 7.4 The release of 'pressures' to reduce disasters: Bangladesh cyclone risk

was heavily affected, reducing employment. In addition food crops and livestock losses were also high (Chung 1987). Subsistence fields were flooded with brackish water. The root crops that had proven themselves more resistant to storm damage over decades had declined in popularity due to the rise of wage income and commercial purchases of food.

Recovery from the hurricane raised many questions about the dependency of the island on a single export crop and dependence of the people on imported food (J.R. Campbell 1984; Pacific Islands Development Program n.d.). Despite appeals for 'modern recovery measures [that] should merge . . . with traditional and time-proven practices' (Chung 1987: 48), there is little evidence that things are moving in this direction.

Rapidly-industrializing coasts

The Philippines is a nation composed of more than 7,000 islands which are home to some 56 million people (J.N. Anderson 1987: 249–50). Much of the long 18,417 km (11,444 mile) coastline is at risk to some of the 200 tropical storms that, on average, occur during a ten-year period in that region of the Pacific. Major population concentrations around the capital, Manila (8 million people), and further north on the heavily populated island of Luzon are frequently affected.

Within one week in October 1989 the Philippines was hit twice by typhoons Angela and Dan which killed 159 people and left 429,000 people homeless (S. Newman 1989). Suffering of this magnitude was recorded nearly every year during the 1980s. For instance, in 1988 there were also two typhoons (Ruby and Skip) which caused extensive flooding in central Luzon and made 200,000 people homeless on Leyte Island.

The cascade of mutually-reinforcing causes for the people's vulnerability begins with land hunger. Land hunger is not the product of overpopulation, but of highly-skewed patterns of land tenure which, despite the Aquino government's commitment to land reform, has changed little (Collins 1989; Boyce 1992). Landlessness leads to a desperate search for livelihood options in the growing cities, where migrants live in poorly-constructed squatter settlements on stilts on the water's edge, in low-lying flood plains and wastelands, and on steep slopes. These are the areas where urban landowners cannot command high rents and where commercial alternatives to shanty development are not profitable.

In 1981, 1.7 million people in greater Manila lived in 415 slum colonies. This is the pattern in another sixteen of the Philippine's cities, where an average of 23 per cent of citizens live in low-income, self-built slums. In total, around 10 per cent of the Philippine population (some 4 million) lived in urban slums in 1981 (Ramos-Jimenez, Chiong-Javier, and Sevilla 1986: 11–13). They remain highly vulnerable to the impact of wind and flooding due to typhoons.

In the rural coastal Philippines, a rapidly-industrializing economy seeks out raw materials, including sites for fish farming and industrial salt production and in the process destroys mangroves (World Resources Institute 1986: 146–50). Land pressure also means that small farmers and indigenous people such as the Palaw'ans (who live on one of the more isolated islands to the west of the archipelago) retreat further into steeper mountain areas. They are forced to do so as government-sponsored migrants and companies take over land at lower elevations for rice farming (Lopez 1987). Clearing land and attempting to farm on such slopes puts these people at risk from the same typhoons that wash away the urban squatter's home. Torrential rains cause deadly landslides such as the one reported in 1988 (Union News 1988; also see Chapter 8). The resulting erosion causes siltation of rivers downstream, adding to the potential for flooding at coastal locations, including the sites of urban slums.

POLICY RESPONSE

Coastal storms demonstrate the dynamics of vulnerability visible in other chapters: the intimate link between vulnerability and livelihood strategies, and the complex and shifting pattern of differential class, ethnic, and age-graded vulnerability. Severe storms also demonstrate vulnerability in some more specific ways. Unprotected, poor-quality urbanization is a process that is basic to vulnerability to wind, flood, and mudslide hazards associated with severe storms. The relationship between town and countryside can exacerbate the hazard when coastal vegetation is removed or disturbed. The role of economic dependency in creating vulnerability emerged clearly from the case of islands dependent on one or a few export crops. Also, we have seen the negative role of absentee landlords in both urban and rural settings.

As with riverine floods, there are a series of policy responses to cyclone hazards. Warning systems can be developed, coastal shelters and meeting points can be constructed, land-use can be regulated by law or financial incentive. Yet severe storms underline the futility of taking only such an engineering-based and administrative approach. It is necessary but not sufficient.

The threat of storm hazard must be placed in the context of 'normal' development policy. This was clear in the case of Andhra Pradesh discussed above. Roads, alternative livelihood possibilities, and availability of credit were highlighted as crucial factors in reducing vulnerability. These are elements of equitable and sustainable development, and are not specifically policies for dealing with disasters. Giving hurricane preparedness in high-risk zones a role as part of mainstream development can also be as straightforward as building community shelters as multiple-purpose structures (e.g. schools), as some NGOs are doing in Bangladesh (Sattaur 1991).

It is essential that we understand in detail the livelihood strategies of all the different social groups affected. Such social profiling is the essential complement to hazard-mapping. In addition, the coping patterns of ordinary people strongly influence their losses and recovery, as well as their survival. So it is vital that evaluation of vulnerability to hurricanes includes detailed accounts of how people cope. The striking differences in recovery rate seen in the case of the Andhra Pradesh cyclone can be avoided if aid is provided in appropriate forms to the more vulnerable groups. This requires an understanding of their coping mechanisms as well as their needs. We noted above that storm-resistant crops are used in Fiji, but did not feature in recovery efforts.

As emphasized so often in this book, there is no substitute for painstaking work *with* the survivors as co-designers and co-directors of the recovery process (Maskrey 1989). In this way it is more likely that vulnerability to the next coastal storm will be reduced. Since rebuilding of houses and other structures and facilities is usually involved in storm recovery, many of these same lessons will be seen in the next chapter, dealing with earthquakes.

NOTES

1 We should also note that, as with floods and some of the effects of volcanoes, there are positive attributes of cyclones for some societies. They can redistribute moisture to drought-prone areas, and are looked on as a positive reduction of drought risk in the Philippines. In Mauritius the authorities regard cyclones as vital to the islands' ecosystem, though of course there are significant dangers involved as well.

2 It is important to note that the valuations of such economic losses may have little relationship with the real losses to livelihood and assets of different types of people. Moreover, comparisons between countries do not take into account the fact that in a poor country the money value of damaged property may be lower, and yet represent a much greater loss in terms of the values of local people.

3 Such indebtedness for boats is a problem facing many small-scale fishing communities, not just those on small islands. Chambers (1983) and Winchester (1986, 1992) have assessed this problem in south India.

4 The Andhra Pradesh cyclone has been analysed in considerable detail in Winchester (1992); much of this section is derived from his work in Winchester (1986).

5 By 1988 5.9 million people were dependent on food aid out of a population of 13 million, many of them forced to flee by RENAMO (Kruks and Wisner 1989: 166; D'Souza 1988).

8

EARTHQUAKES, VOLCANOES, AND LANDSLIDES

INTRODUCTION

Earthquakes, volcanic eruptions, and landslides, for all their dramatic impact, do not remotely match the scale of casualties that result from droughts, floods, and coastal storms (Sapir and Lechat 1986). This century to the end of 1990 there have been an estimated 1.52 million officially reported deaths from earthquakes. Almost half this total have occurred in China, which also suffered the most devastating single event in the 1976 Tangshan earthquake which resulted in 242,000 deaths (Coburn and Spence 1992).

The seventeen most severe volcanic eruptions of this century have resulted in 75,000 deaths, with the most catastrophic eruption occurring in Mount Pelee in Martinique in 1902 when 29,000 were killed, and the second most severe event being the eruption of Nevado del Ruiz in Colombia in 1985 with the loss of 23,000 lives. Thus the remaining fifteen volcanic eruptions averaged 1,582 deaths per event (Wood 1986; United Nations 1985).

In the case of landslides, forty sudden-impact landslides have been reported this century, causing 271,072 deaths. However, this includes the most damaging landslide of this century, which took place in Gansu province, China in 1920, when 200,000 were reported killed. In 50 per cent of these disasters fewer than 100 were killed (Alexander 1989). Thus it can be seen that in global terms, landslides have relatively low casualty statistics relative to other hazards. However, the data are misleading, since landslides often occur as a secondary consequence of another type of hazard, such as flooding, a cyclonic storm, or as a result of an earthquake. So landslide casualties are often added to the total deaths and injuries attributable to these broader events, and those specifically linked to landslides are probably underreported.

VULNERABILITY TO EARTHQUAKES

The effects and consequences of earthquakes are varied, but a key issue is the relationship of earthquakes to unsafe structures (Cuny 1983; Coburn and

Spence 1992). From the evidence of past disasters, it is clear that many countries in seismic areas, particularly in developing countries, possess many highly dangerous structures that may collapse under extreme seismic forces (French 1989). In some instances they may be so dangerous that they can even collapse of their own accord without the assistance of unusual forces.

Over 95 per cent of all deaths in earthquakes result from building failures (Alexander 1985). Seaman has commented as follows on the relationship between mortality and buildings: 'Variations in mortality among different countries are primarily due to differences in building styles and density of settlements. The overwhelming majority of people who die in earthquakes are killed by the collapse of man-made structures, particularly domestic dwellings' (Seaman, Leivesley, and Hogg 1984: 10–11). Seaman then identifies four critical variables:

(a) the seismic and geologic features of an area, the building design and construction, and specific aspects of building construction and the risks to the occupants; (b) the location of the inhabitants (e.g. indoors or outdoors); (c) the age and sex of inhabitants and those killed or injured; (d) the types of injury, severity and timing of presentation for treatment.

(ibid.)

It is possible to add a further item to this list that relates to the timing of earthquakes. If they occur during the night (as the 1976 Guatemala earthquake), then casualties are always higher for two reasons. Firstly, people are likely to sleep through the foreshocks which in daytime enable more time to escape from buildings. Secondly, when lying flat in bed the body is highly exposed to injury from falling debris.

Since buildings are such critical factors in seismic risk, concerned officials simply need to look very hard at the elements that relate to safety with specific consideration of the shape, siting, and construction details of buildings to find answers to three vital questions. Firstly, where are buildings likely to fail? Secondly and perhaps even more significantly, what are the root causes of this dangerous situation? Thirdly, what action can be taken to reduce this underlying pressure?

If the first question is tackled while the second is ignored, then the lives and property of the population will remain at risk even though certain individual buildings are made safe, since symptoms (unsafe building design) will have been addressed rather than the root causes.

Cardona and Sarmiento (1990: 22) have also examined the issue of the vulnerability of communities to disasters with a specific focus on their health status. They suggest considering the following ten variables (p. 172):

169

Box 8.1 The Guatemala earthquake, 4 February 1976

This earthquake was a crucial experience for many agencies involved in disaster assistance in both negative and positive terms. Some major 'aid blunders' were committed and some innovative ideas in the education of small builders to construct earthquake resistant (aseismic) houses were pioneered (Cuny 1983: 164–93).

The disaster focused attention on the urban and rural poor people's vulnerability to exploitation by the landlords of Guatemala, and on certain aid agencies which pursued assistance policies they probably now regret, or acknowledge as important 'learning experiences'.

The earthquake killed 22,000 people living in unsafe houses in the rural highlands of Guatemala as well as within dangerous squatter settlements in Guatemala City. It left the upper and middle classes virtually unscathed. This was the first major earthquake widely recognized as having had such a markedly selective impact, hence its designation by an American journalist as a 'class-quake'.

Vulnerability variations can be clearly detected in the Guatemalan case. Firstly, there was a strong ethnic as well as class factor at work. The highland rural people who died were not only poor, but were indigenous Mayan Indians. The dead in Guatemala City (some 1,200 people) and the 90,000 made homeless were almost exclusively concentrated in the city's slums (*Latin America*, 9 April 1976: 115). Secondly, it was exceedingly difficult for either Indians or urban squatters to obtain post-disaster assistance from the government.

The socio-economic forces that led to so many people living in unsafe conditions, and the political forces that controlled post-disaster aid, were a mirror of the society at large (Plant 1978). What made Guatemala unusual was the high degree of awareness of these social weaknesses on the part of a large proportion of the population, so that post-disaster relief and rehabilitation became a political battleground. In the words of a contemporary journalist:

> In this well-known fault zone the houses of the rich have been built to costly anti-earthquake specifications. Most of the poorest housing, on the other hand, is in the ravines or gorges which are highly susceptible to landslides whenever earth movements occur. The city received proportionately little aid largely because it is governed by the most radical opposition tolerated in Guatemala, the *Frente Unido de la Revolucion*, a social democratic coalition. Its leader Manual Colon Arguetta, was wounded by unknown gunmen on 29 March, the latest victim of a wave of terror attacks that has claimed 40 lives since the earthquake. One city official, Rolando Andrade Peña, was shot down two weeks after the earthquake after suggesting that homeless people should be encouraged to rebuild on unoccupied private land.
>
> (*Latin America*, 9 April 1976: 115)

In 1989, thirteen years after the earthquake, one of the present authors revisited Guatemala City to determine whether the people were any less vulnerable than they had been in 1976. In many ways matters now seem more positive. While there are still houses on the steep slopes, they are certainly not as congested or precarious. Many of the urban poor who lost their homes yet managed to survive the earthquake immediately vacated the most dangerous slopes for flat or gently sloping sites a short distance away. This illegal 'invasion' took place from the day of the earthquake onwards, and ever since the barrio has been known as '4th of February'. When survivors first 'invaded' safer sites, there were a large number of visiting newsmen in the city to report on the disaster and the authorities turned a blind eye to the influx of displaced families. Eventually, perhaps due to the sheer force of numbers linked to sustained political pressure, the occupiers were granted legal titles to the land by the government.

However, there is no evidence that the builders of these houses had any knowledge of earthquake-resistant construction. So although their sites are safer from earthquake-induced landslides, flash-floods, and eviction orders, their dwellings remain dangerous. In fact the risk of their houses collapsing may have significantly increased. When they were illegally sited they were generally built out of lightweight materials, including corrugated iron sheet roofing, but when they were legalized many families began to build in heavy materials such as reinforced concrete which is likely to cause greater damage than structures built of lighter materials.

Also, while there is some evidence of progress in Guatemala City, there remains a very depressing picture of political repression linked to reconstruction activity in the rural highlands of Guatemala. There, in the early 1980s, tens of thousands of highland Indians were killed by the military in disputes over expropriation of Indian land (see Chapter 9).

Oxfam America was one of the many NGOs that was heavily involved in the reconstruction programmes based on co-operative activity. In 1982 they published an account of the reign of terror that ensued, including a series of interviews with local leaders:

> 'the earthquake tore open many holes in the social fabric which had already been stretched thin. The rich and those in power came out richer and the poor came out poorer, and differences and inequalities became more visible. More protest led to more repression to contain the forces of change. Those in power do not want to share the wealth'. 'I think this region has become the target for increased repression and violence against the population . . . (since) many people in this area were very active in reconstruction efforts after the earthquake.'
>
> (Quoted in Davis and Hodson 1982: 15)

Miculax and Schramm wrote a case-study of the long-term consequences of one of the Housing Education programmes in 1989, thirteen years after the earthquake:

> A terribly unfortunate negative consequence of these improvements in community organisation should be noted. During the 'violence' of the 1980s, individuals who had developed their personal capacities during the post-disaster relief project were seen as 'troublemakers'. Many were killed by the army and others sought exile in neighbouring countries.
>
> (See Anderson and Woodrow 1989: 237)

In Guatemala 'political vulnerability' expanded as a direct consequence of community development and leadership training specifically intended to reduce vulnerability to economic factors or seismic hazards.

Age structure;
Health structure–morbidity;
Health structure–mortality;
Family income;
Illiteracy rate;
Level of schooling;
Location of the workplace;
Spatial distribution of the population;
Urban population density;
Rural population density.

To investigate the root causes of vulnerability to earthquakes more carefully, we consider two widely differing earthquake-related disasters in Central America: Guatemala in 1976 (Box 8.1) and Mexico City in 1985 (Box 8.2). Not only were the patterns of vulnerability quite different, but also the recovery operations were of a very distinct character.

Vulnerability to hazard warnings

Before leaving the subject of earthquakes, it is important to consider the issue of prediction. Is this a developing science that will greatly reduce casualty patterns, as is already beginning to happen in the case of cyclone deaths, or is it a false trail that may be a contributor to new forms of vulnerability? The predictive processes available for some other hazards are not yet sufficiently developed to forecast earthquakes. However, there has been excellent progress in the monitoring of landslides and volcanic risks. Already Hong Kong has sophisticated computer models in routine operation to monitor rainfall levels and predict when a given slope is likely to collapse. Using the technique the authorities have made a number of successful evacuations that have preceded major landslides (Whitcomb 1990). Several successful evacuations of vulnerable communities have taken place in recent years as a result of effective warnings of impending volcanic eruptions (United Nations 1985). Examples of such successes include the eruptions of Mount Pinatubo in the Philippines, Etna in Italy, and Mount St Helens in the US.

The evacuation that preceded the main eruption of Pinatubo was a result of timely action by the authorities and the pressure of events. Pinatubo was regarded as extinct, so there was understandably minimal planning for a possible eruption (International Federation of Red Cross and Red Crescent Societies and Centre for Research in the Epidemiology of Disasters 1993: 62; Newhall 1993; Baxter 1993; Tayag n.d.(b)). Volcanic activity began on 2 April 1991, and the residents on the mountain slopes within the likely affected area were encouraged to evacuate with an incentive offer of relief supplies in the emergency evacuation centres. Initially the response was slow, with about 25,000 leaving their homes, but this changed after 9 June when the first major eruption occurred. The government and the Philippine National Red Cross created 276 evacuation centres that ultimately housed 130,944 people. On 15 June the area suffered a second disaster as a typhoon devastated the region.

There is no doubt that the volcanologists gave wise advice to the National Disaster Co-ordinating Council (NDCC) who ordered the evacuation, but it was probably the further eruptions that persuaded most people to evacuate. As a result, casualties of 321 deaths and 275 injuries were very low (these figures include casualties from subsequent mudslides). However, there were also serious negative effects of the evacuation that should be considered. Many sick and elderly people died in the evacuation areas. Communal living in camp conditions, with cyclonic rainfall to compound volcanic ashfalls, was a very unhealthy living environment (Newhall 1993; Baxter 1993). This problem was probably unavoidable then, though more attention needs to be given to evacuation procedures.

Hazard-mapping will reveal the location and probable severity of earthquakes and landslides. Long-term predictions, based on a 'gap-theory' about the anticipated general location of a forthcoming earthquake may also be useful. But the current state of knowledge gives no precise warning of their timing. It would appear that this lack of earthquake warning is a significant factor in maintaining the vulnerability of people.

Early warning of drought, cyclone, and flood is already reducing the vulnerability of communities previously at risk and significantly reducing deaths and injuries. But an earthquake prediction can result in other problems, such as panic in evacuation, or the more mundane issue of blighted property and falling values in areas thought to be targets. The risk of legal action over such issues can actually make prediction politically unacceptable. In 1986, following small tremors, the population of 56,000 in the towns of Lucca and Modena in Tuscany (Italy) were evacuated. Thirteen thousand hospital beds were made available, and railway carriages brought in to accommodate the evacuees. Widespread traffic jams and petrol shortages followed, and shops and businesses closed for two days. However, no earthquake occurred and recriminations from angry businessmen eventually caused the resignation of the mayor and administration (Coburn and Spence 1992: 41–2).

Many commentators have suggested that a total evacuation from a major city would constitute a disaster in its own right, perhaps of far greater impact than any impending earthquake. They have surmised that there would be traffic problems and accidents on a Herculean scale (particularly if there was widespread panic), crime and looting, economic losses beyond belief, and acute difficulty in maintaining public services for the metropolitan region and perhaps for an entire nation. Additionally, there would be public health risks to the displaced population, depending on where, how, and for what duration they were accommodated. These would include epidemics due to inadequate sanitation, the psychiatric stress of uncertainty, anxiety, and

Box 8.2   The Mexico City earthquake, 19 September 1985

The impact of this disaster was very different to Guatemala. While there are millions of people living in the 'informal settlements' of Mexico City, in conditions very similar to the steep ravines of Guatemala City, they were not the victims of this earthquake. Those who suffered in Mexico City were not the poorest residents, and this case reminds us that vulnerability is not identical with poverty, although the two are often strongly associated. We will need to analyse the situation in several ways in order to understand all the relationships that determined vulnerability in this complicated case. We will use the 'pressure and release' model developed in Chapter 2. However, the origins of vulnerability have such a long history, extending back to the conquest by the Spanish, that we will have to inspect past 'layers' as well.

Within Mexico City, currently the largest urban conurbation in the world with over 19 million residents, the long-suffering population have had to learn to cope with a host of natural and technological hazards. Earthquake remains a major threat and triggered the disaster that led to 5,000 deaths in 1985. But it is not the risk which is highest on Mexico's political agenda, which in the 1990s is atmospheric pollution. There are other natural and man-made hazards, including the risks of flood, landslide, earthquake, ground subsidence, fire, traffic congestion, and hazardous factories, to name some of the environmental problems. To compound the problem many of these hazards are interactive.

If a comprehensive process of risk assessment involving hazard-mapping and vulnerability analysis were to be carried out in Mexico City, then all natural and technological hazards and patterns of vulnerability would need to be synthesized into a comprehensive planning tool. That tool has yet to be developed. However, for purposes of a much narrower analysis of earthquake risk, a series of 'vulnerability strata' will be identified. These begin with the 'physical environment', then move into the 'urban fabric', followed by the 'society' living and working within the buildings, and end with a consideration of the 'economic activity' of the affected community. This will explain some of the historical aspects of seismic risk, as well as the complex multidisciplinary nature of vulnerability analysis.

First layer: historical influences on the physical environment

The seismic hazards facing Mexico City can be mapped with some accuracy in terms of their frequency, severity of impact, damage patterns, type of ground motion, and location in relation to topography and soil conditions.

However, even such physical factors can be affected by human intervention. Tobriner has written an account of the way Mexico City has developed over the past six centuries on a site which could hardly be more dangerous. Over this period it has been at risk from floods, soil shrinkage, volcanic activity, and earthquake impact. He observes that it is 'one of the great ironies of urban history that Mexico City, perhaps the largest city in the world, stands on one of the planet's most unstable soils' (Tobriner 1988: 469–79). Figure 8.1 shows the relationship between building damage and the bed of the old Lake Texcoco, with its legacy of these dangerous conditions. The explanation of why this occurred is rooted in the history of the site from the thirteenth century, when the Aztecs made it their capital, Tenochtitlan. As Oliver-Smith points out in his account of the '500 hundred year earthquake' in Peru, the root causes of vulnerability sometimes lie in the very remote past.[1] When the Spanish arrived they found it a well-adapted settlement. But conquest and its symbolic needs required them to destroy it and substitute a new city which required the draining of Lake Texcoco.

However, the site of Tenochtitlan suffered from four natural hazards: volcanic eruptions, earthquakes, drought, and severe flooding. The historic centre of Mexico City now sits on a lake-bed, with an alluvial subsoil up to 60 metres (200 feet) deep. In the 1985 earthquake this soil behaved like a liquid, with massive ground shaking causing damage almost exclusively within the area of the original lake-bed. This is the tragic legacy of a major city set on an unstable site which has its roots in the power of Aztec kings and colonial rulers a full four and a half centuries ago. In Figure 8.2, below, this preconditioning factor is depicted as a 'root cause' of vulnerability.

Second layer: buildings at risk

The next layer of vulnerability concerns the buildings set on this lake-bed subsoil. They are particularly vulnerable to two types of risk. The first is of ground-shrinkage, resulting in the subsidence of buildings on the alluvial soils. This is happening because the water-table is dropping and the clay drying out owing to excessive extraction of water for the city's needs. This is a form of continual 'creeping' or pervasive disaster that can cause extensive damage to property but no direct loss of life. The second more severe type of risk is that of earthquake impact, likely to cause extensive casualties and damage to property.

In an assessment of building vulnerability to seismic risk, a survey has been made of about 20,000 buildings in the historic centre. Seventeen factors are considered in the survey, including such matters as: levels of maintenance, the shape of buildings, the 'hammer effect' of one building knocking another during an earthquake, building height, and type of construction (Aysan *et al.* 1989).

One major factor that has still to be considered in mapping seismic risk concerns the age of all reinforced concrete buildings. Those erected between 1925 and 1942 were built to very high standards of workmanship. Then from 1942 to 1964 the quality was very poor due to a building boom and a consequent lack of supervision. From 1964 to the present, seismic codes have applied, and the quality of workmanship has improved (Ambraseys 1988).

However, the detailed analysis of earthquake damage has revealed that the primary factors that caused damage to buildings related to their siting and their height. It was found that on sites within the lake bed, rigid structures (such as stone masonry buildings) generally performed better than flexible ones (such as reinforced concrete structures).

Height was an even more significant factor in vulnerability, where medium- to high-rise buildings between six and twenty storeys were worst affected, with particularly severe damage to buildings of between nine and eleven storeys high. The reason for this phenomenon relates to the sensitivity of

Box 8.2 continued

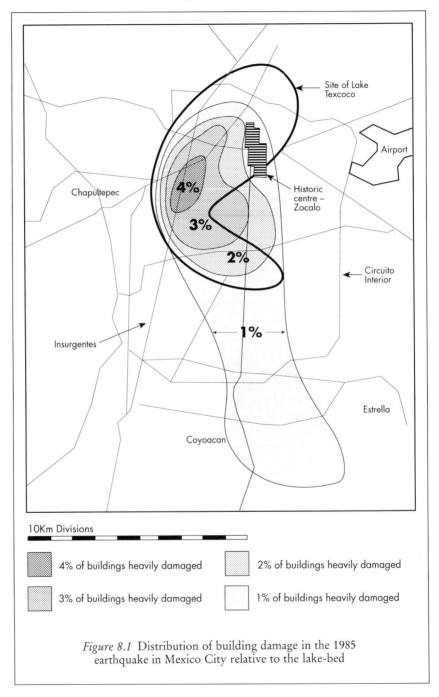

Figure 8.1 Distribution of building damage in the 1985
earthquake in Mexico City relative to the lake-bed

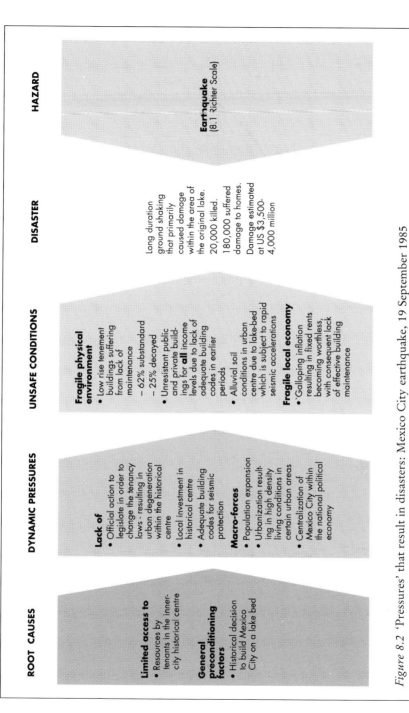

ROOT CAUSES	DYNAMIC PRESSURES	UNSAFE CONDITIONS	DISASTER	HAZARD

Limited access to
 • Resources by tenants in the inner-city historical centre
 	
 General preconditioning factors
 • Historical decision to build Mexico City on a lake bed
	Lack of
	 • Official action to legislate in order to change the tenancy laws - resulting in urban degeneration within the historical centre
	 • Local investment in historical centre
	 • Adequate building codes for seismic protection
	
	 Macro-forces
	 • Population expansion
	 • Urbanization resulting in high density living conditions in certain urban areas
	 • Centralization of Mexico City within the national political economy
		Fragile physical environment
		 • Low rise tenement buildings suffering from lack of maintenance
		   – 62% substandard
		   – 25% decayed
		 • Unresistant public and private buildings for **all** income levels due to lack of adequate building codes in earlier periods
		 • Alluvial soil conditions in urban centre due to lake-bed which is subject to rapid seismic accelerations
		
		 Fragile local economy
		 • Galloping inflation resulting in fixed rents becoming worthless, with consequent lack of effective building maintenance
			Long duration ground shaking that primarily caused damage within the area of the original lake.
			 20,000 killed.
			 180,000 suffered damage to homes.
			 Damage estimated at US $3,500-4,000 million
				Earthquake
				 (8.1 Richter Scale)

Figure 8.2 'Pressures' that result in disasters: Mexico City earthquake, 19 September 1985

Box 8.2 continued

Box 8.2 continued

high-rise buildings, with their lower natural frequencies of vibration than low-rise buildings. They were prone to resonate with the low frequency seismic energy that emanated from the epicentre 370 km (230 miles) away (Degg 1989).

A very serious characteristic of the disaster was the destruction and damage to public buildings and those with high levels of occupancy combined with constant use such as hotels and hospitals. Five hospitals collapsed and twenty-two were seriously damaged which resulted in a 28 per cent loss of public hospital capacity at the precise time when it was most needed. In the collapse of just six buildings 1,619 people lost their lives (Kreimer and Echeverria 1991).

In addition to the failure of large high-rise buildings there was another group of buildings that suffered severe damage which was not as well publicized. These were smaller structures, with high levels of occupancy, with mixed commercial and domestic use. Analysis of casualty statistics clearly indicates the rather obvious fact that in the event of an earthquake occurring during daytime, people manage to escape from low-rise buildings more easily than from high-rise structures (Aysan et al. 1989).

When looking at the building damage in Mexico City in relation to the 'pressure and release' model, it is clear that the cause and effect process is not as straightforward as Guatemala. In the Mexico case, when considering the damage to high-rise structures, there are no easily identified underlying causes other than the lack of engineering knowledge. It is possible to see some relationship between underlying causes and unsafe conditions in the older tenements, where there has been a failure to maintain severely overcrowded buildings. In some instances there was evidence of failure to supervise building construction adequately. Such failures are often practical or economic as well as an ethical matter.

The question of the level of building maintenance relates to patterns of ownership and occupation as well as the role of the state in enforcing maintenance standards. These issues provide a link with the next layer of the study, that of human vulnerability.

Third layer: society at risk

Everything discussed in the previous layers represents well-established areas for the physical mapping of hazards. The two layers that follow are fields of risk assessment still in their infancy, and their precise linkage to physical hazards must be defined in specific terms.

In contrast to physical hazard-mapping, human vulnerability analysis covers a bewildering diversity of topics that concern social patterns and institutions (termed 'structures of domination' in the access model), society-wide and intra-household social relations, economic activity (gender and age relations are particularly important), and the psychology of risk.

Given the distribution in time and space of the earthquake event (hazard) itself and the distribution, just described, of historically remote 'causes' and unsafe buildings, what specific mechanisms (dynamic translating processes) were at work in 1985 that placed certain people in those unsafe buildings at the critical moment? These include the density of population (a function of equitable location factors such as site of employment, land prices, rents), the ownership of buildings relative to their maintenance, patterns of building use (seen in terms of both space and time), the perception of risk of the local population, cultural values such as desire to remain in one's natal neighbourhood, and the existence of local institutions that could play a key role in post-disaster recovery.

Mention of the last 'mechanism' that translates 'cause' into vulnerability reminds us that disasters relate to the long-term consequences for survivors as well as the immediate impact. This was clear in the discussion of recovery between cyclones in Andhra Pradesh (Chapter 7). In the context of the Mexico City earthquake, there are several critical issues concerning recovery. These include such economic matters as the capacity of people working in the high-risk zone to restore their livelihoods to their pre-disaster level following another earthquake. Much of these data are concerned with potential economic losses which are considered as a final layer of vulnerability.

Fourth layer: the local economy at risk

Effective disaster planning has to consider the likely effects on the local economy of earthquake losses. These data, which did not exist before the Mexico City earthquake, can be measured in three ways. Firstly, there are direct losses (e.g. of a building or factory in a future disaster), or secondary losses (e.g. fire damage caused by the earthquake), or indirect losses (e.g. the loss of income as a result of the local population not being able to purchase goods due to their temporary loss of income, or because of interruptions to supplies).

The 'layers' of Mexico City's vulnerability to earthquakes provide a convenient 'access model' for the examination of earthquake hazards. The disaster did not occur in a vacuum, its impact was suffered within a 'space–time' context that affected a particular section of the city's building stock and population occupying specific structures at a specific time.

The effect of the earthquake can be considered in two ways: the destruction of property and the impact on lives. In the Mexico City earthquake, two distinct categories of buildings collapsed or were damaged; both were constructed on alluvial soils that formed the bed of a long-vanished lake. The first category involved people who died or were injured in high-rise buildings, including a hotel and a number of hospitals. These casualties came from all levels of Mexican society. In contrast, the second group were the predominantly low-income residents of nineteenth-century low-rise tenements. As we pointed out in Part I, we are concerned to define and to underscore the relevance and practicability of vulnerability analysis by pointing out how it is applicable beyond simple criteria such as income and status. The losses in the first category included some victims who would not be considered vulnerable in terms of their income. However, vulnerability is closely parallel to low-income status of residents of the tenements.

The earthquake occurred at 7.00 a.m. on 19 September 1985 when most people were on their way to work. For those on foot, the hazard was falling

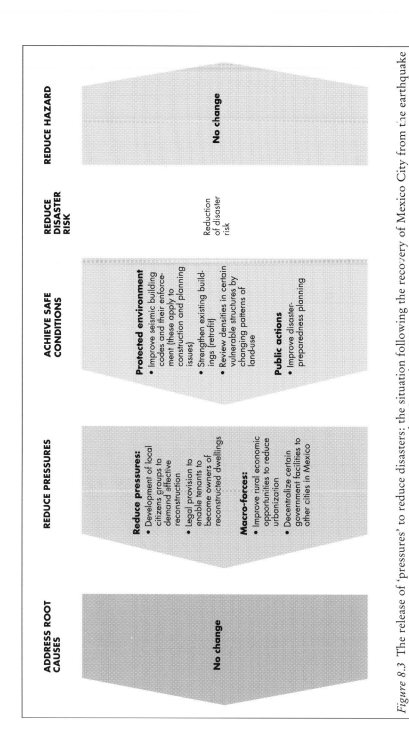

ADDRESS ROOT CAUSES

No change

REDUCE PRESSURES

Reduce pressures:
- Development of local citizens groups to demand affective reconstruction
- Legal provision to enable tenants to become owners of reconstructed dwellings

Macro-forces:
- Improve rural economic opportunities to reduce urbanization
- Decentralize certain government facilities to other cities in Mexico

ACHIEVE SAFE CONDITIONS

Protected environment
- Improve seismic building codes and their enforcement (these apply to construction and planning issues)
- Strengthen existing buildings (retrofit)
- Review densities in certain vulnerable structures by changing patterns of land-use

Public actions
- Improve disaster-preparedness planning

REDUCE DISASTER RISK

Reduction of disaster risk

REDUCE HAZARD

No change

Figure 8.3 The release of 'pressures' to reduce disasters: the situation following the recovery of Mexico City from the earthquake of 19 September 1985

Box 8.2 continued

Box 8.2 continued

masonry crashing onto pavements, but for the thousands inside metro trains or motor vehicles the immediate environment was highly protected.

The vast majority of those who died were inside medium- to high-rise buildings, in the central area of the city. Some 12,700 buildings were affected, 65 per cent of which were residential. The housing for 180,000 people was damaged, and 50,000 needed temporary accommodation (Kreimer and Echeverria 1991). Since the damage affected high-investment buildings, the financial losses were enormous, estimated at US$4,000 million (Kreimer and Echeverria 1991). The reinsurance industry has assessed the earthquake as one of the three most disastrous of this century, the others being the San Francisco and Tokyo earthquakes of 1906 and 1923 respectively (Degg 1989).

Such financial loss dwarfs the dollar value of dislocated livelihoods. Yet for those who relied on work in the 1,200 small industrial workshops destroyed, the cost was great. Once more it is clear that issues of recovery and rehabilitation cannot be separated from the profile of vulnerability. Are those workers still unemployed, or have they found an alternative? Did the earthquake begin a spiral into poverty for those households?

The number of casualties varies in spatial terms, since it relates to the horizontal location of victims (i.e., in a building set on the alluvial soils), or the vertical location (at a certain height in a building). The timing of the earthquake was even more critical in determining the number who died or were injured. Had an earthquake with identical characteristics and intensity occurred just 3 hours earlier, when people were asleep, there would have been a significant increase in casualties, although property losses would not have changed.

Figure 8.3 is an attempt to identify the factors which could be adopted by the Mexican authorities to release the pressures that have in the past created unsafe conditions in Mexico City (Coburn and Spence 1992: 130; Gomez 1991: 56–7; Echeverria 1991: 60–1). The model addresses vulnerability indirectly and directly. Indirect measures include reducing the size of Mexico City. Urbanization is addressed by improving rural economic opportunities, lessening the migration to the city, and by decentralizing some of the federal government's functions (hence employment opportunities) to other cities in Mexico. In this way one of the most important dynamic pressures that translate global root causes into unsafe conditions is relieved.

In addition, unsafe conditions are addressed directly by improved aseismic building codes and their enforcement, strengthening existing structures, and reducing the densities in certain weaker structures by changing patterns of use. These steps, none of them unthinkable in contemporary Mexico, although some are more difficult politically and more expensive than others, complement a programme of improved disaster-preparedness planning. But already, less than ten years after this major disaster, there is a decline in 'political will' to protect the city from such disaster impact. Seismic risks have been displaced by other more pressing political and environmental problems.[2]

possible separation from loved ones. These issues will probably weigh most heavily with governments facing political instability.

While governments may tremble at the prospect of a major evacuation, the prospects are even more devastating when seen at the 'micro-level' of poor families with very limited resources. Wisner describes how they are

> hurt when they have to miss a few days or weeks of work, often the price of temporary relocation or evacuation. These people are not 'stupid'. The mental arithmetic itself is cruel: evacuation on a 'false alarm' could make the difference between life on the poverty line with some hope and years of absolute poverty below it, because of property theft by looters, loss of work, or damage to the unprotected home.
>
> (1985: 16)

Such a critique of evacuation options may be regarded as decidedly 'doom-laden', but there is mounting evidence of the problems that mass evacuations pose for authorities. An epic example was the management of several million Kurdish refugees from Iraq in the spring of 1991 or speculation by the authorities of what would happen if the entire length of Miami Beach Island had to be evacuated in the event of a major impending hurricane.

Tremendous social and economic problems have resulted when entire regions have been evacuated due to volcanic risk. A classic case is the Caribbean island of Guadeloupe in the Lesser Antilles, where 73,000 people were evacuated in 1976 from the high-risk zone for three and a half months, resulting in huge economic losses and great social strain on the population and their government. The volcano never erupted and there was only minor volcanic activity (Blong 1984).

LANDSLIDES

Landslides involve the movement of material that may vary considerably in its character, including rock, debris, mud, soil, or several of these in combination (Alexander 1989: 157). Alexander includes landslides that are generated by a wide variety of 'agents': the failure of coal-mining waste in Wales (Aberfan), a dam burst in Italy (Vajont), a volcanic eruption in Colombia (Nevado del Ruiz), an earthquake in Peru (Mt Huascaran), and flooding in Brazil (Rio de Janeiro).

In applying vulnerability analysis to the case of landslides, we need to move beyond the physical hazard to inquire about human activities that might act as 'triggers' for the physical event (e.g. the location of a dam) as well as the manner in which people are exposed to the risk according to their varying characteristics. Differential ability to recover after a landslide is also important since it can cause people to be more exposed to future risks, as we have seen in previous chapters.

Consider four typical examples of landslides that took place between 1985 and 1988:

1 Mameyes, near Ponce in Puerto Rico, 8 October 1985, killing 180, 260 homes destroyed (Wisner 1985; Doerner 1985);

2 Rio de Janeiro, Brazil, February 1988, where 277 were killed, 735 injured, and more than 22,000 displaced in shanty towns (Allen 1994; Byrne 1988; Margolis 1988; Michaels 1988; Munasinghe, Menezes, and Preece 1991: 28–31);

3 Catak, Turkey, on 23 June 1988, killing approximately seventy-five (Gurdilek 1988);

4 Hat Yai, Thailand in November 1988, killing 400 (*The Economist* 1989; Nuguid 1990; West 1989).

The analysis of the likely causes (or 'general predisposing factors' in terms of the PAR model) of these landslides shows a number of interesting similarities.

The first commonly-cited cause is deforestation. There was an outcry against logging in Thailand following the landslides. West notes that this protest

> did not come from bearded ecologists and trendy 'green' politicians, but from the local farmers and townspeople, those in fact, who had suffered. The anger comes from below, and is aimed especially at the greedy loggers, frequently Chinese businessmen in partnership with senior officials in the police and army.
>
> (1989: 18)

The Prime Minister of Thailand visited the site of the disaster and announced that logging operations would be banned. Forty years ago, 70 per cent of the country was covered by forests, but in 1989 this percentage had dropped to just 12 per cent (*The Economist* 1989).

In Rio, the authorities were criticized for not taking effective action to tackle the problems of the denuded hills where all the *favelas* (squatter settlements) had been constructed. These housed a million people out of the 8 million in the city. There had been extensive deforestation in these areas to make way for dwellings as well as providing fuel wood. Socio-economic factors are an obvious 'pressure' that both forces squatters to inhabit unsafe locations and forces them to cut vegetation for fuel or building material since alternatives are too expensive (Allen 1994).

Poorly located road-building is also commonly mentioned as a cause. The Turkish authorities commented that the roads in Catak should have been cut into the contours, rather than running parallel with them. Frequently, roads are cut into steep slopes with minimal understanding of the geomorphology of the setting, and can interrupt drainage patterns. The actual 'cut and fill technique' of road-building on steep slopes can contribute to landslide risk (K. Smith 1992: 165).

Environmental damage to subsoil stability is also frequently cited as a cause. Changes in the water-table can occur due to leaking tube wells, standpipes, and septic tanks, and appear to have been a contributory cause of the landslides in Puerto Rico and Rio de Janeiro. Unsafe, unauthorized building on dangerously steep slopes is very often cited as a cause of landslide disasters. The location of squatter settlements themselves may have been a contributory cause to the landslides in Puerto Rico and in Rio de Janeiro.

Warning systems for predicting water flow and arranging the evacuation of communities at risk are often lacking in urban areas. This appears to have contributed to the landslide disasters in Puerto Rico and Rio de Janeiro. There had been extensive rain for a number of days prior to the mudslides, but no monitoring or advance planning for such contingencies.

VOLCANOES AND RELATED HAZARDS

Volcanoes are vents in the crust of the earth through which molten rock is extruded as lava or ejected as ash or coarser debris, sometimes accompanied by steam, and hot and often poisonous gases (Davis and Gupta 1991: 29). Associated hazards include earthquakes, and mud and rockslides. Volcanoes are like some epidemics (discussed in Chapter 5) in that they represent a limit to the use of vulnerability analysis. Volcanic eruptions endanger any person living within the high-risk zone, whether rich or poor, landowner or landless farm labourer, man or woman, old or young, member of ethnic minority or majority. Tomblin has commented: 'Eruptions differ from most other major causes of disaster such as earthquakes, hurricanes and floods, in that they cause virtually total destruction of life and property within relatively small areas which can be easily delineated' (1987: 17).[3] Poisonous gas emissions do not differentiate between social groups. But even where these are not the main threat, income levels, the quality of house construction, and the type of occupation all seem to have little bearing on people's differential capacity to resist the volcanic arsenal of hot gas emissions, blast impact, lava flows, projectiles, volcanic mudslides (*lahars*), and the deposit of ash.

It may be argued that wealthy people have more access to knowledge, which can include an awareness of volcanic risk, and therefore they are better able to respond to warnings to evacuate in the event of a likely eruption. But there is growing evidence that poor people living near active volcanoes are aware of the risks. Once they observe signs of volcanic activity they are just are as likely to follow evacuation orders as their rich neighbours (Kuester and Forsyth 1985; Tayag n.d.(b); Zarco 1985).

The term 'hazard' is not strictly accurate since in many cases they bring major benefits as well as havoc: irrigation and fertile silt from flooding, or the rainfall over drought-prone land from tropical cyclones. This process is probably better seen in the case of volcanoes than any other geological hazard, since there are no obvious benefits from landslides and earthquakes.

The products of volcanoes can be highly beneficial to any society, and include extremely fertile soils resulting from the weathering of volcanic ashes and pyroclastic materials. Farmers often obtain bumper harvests as a result of a mild sprinkling of volcanic ash on their fields (Wood 1986: 130). In April 1992 Cerro Negro erupted near Leon in Nicaragua. A thick layer of volcanic ash was deposited, with gloomy forecasts that the agricultural economy would be interrupted for years. However, within ten months farmers were already enjoying good crops from the fertile soils intermingled with volcanic ash (Baxter 1993).[4]

Such volcanic blessings undoubtedly constitute an extremely powerful social and economic magnet. It is often suggested that people inhabiting high-risk zones are gamblers by nature, who take big risks to achieve uncertain benefits. But the odds are very uneven since it would not appear to take families very long to decide to face the risk of an eruption with a return period of perhaps forty-five years, for the pay-off is significantly enhanced economic opportunities that will apply every day. Paradoxically, effective disaster-preparedness with its expectations of good warning and evacuation by the authorities only adds to the power of the magnet drawing people into high-risk zones (see Box 8.3).

POLICY RESPONSE AND MITIGATION

In concluding this chapter on a positive note, four approaches to risk-reduction in the face of geological hazards are suggested (we go into these in more detail in Chapter 10).

Firstly, earthquake, landslide, and volcanic disasters can be used to change unjust structures. Popular development organizations can capitalize on a disaster event to challenge and possibly change vulnerable, unjust political, social, and economic structures. Holloway has suggested that

> Disasters will often set up a dynamic in which social structures can be overturned, and relief and rehabilitation judiciously applied can help change the status quo; while projects will be the models in microcosm that can be used to demonstrate to government the possibilities of a variety of ways of working.

(1989: 220)

Thus in the aftermath of the Mexico City earthquake, neighbourhood organizations were strengthened and increased their demands for government services (Robinson *et al.* 1986; Annis 1988). There is not a direct relationship between the strength of local organizations and reduction of vulnerability to disaster, but certainly the converse is true: in the absence of grass-roots and neighbourhood organization, vulnerability increases.

Box 8.3   Case-study: pre-disaster planning, Taal volcano, the Philippines

An example of the conflicting demands of prosperity versus safety and how they relate to public policy can be seen in a case-study of Taal volcano in the Philippines. This is one of the world's lowest and deadliest volcanoes, on an island in Lake Taal, about 60 km (40 miles) south of Manila. Taal has had thirty-three recorded eruptions since its earliest recorded outburst of 1572. The 1911 eruption resulted in 1,334 deaths and covered an area of 2,000 sq km (770 sq miles) with ash and volcanic debris which fell as far away as Manila. Further volcanic activity occurred in 1965, 1966, 1967, 1968, 1970, 1976, and 1977 (Arante and Daag n.d.; Philippine Institute of Volcanology and Seismology n.d. (a)). The main eruption of 1965 was completely unexpected, and no official warning was issued. Chaos ensued, as Arante and Daag describe:

> panic gripped the beleaguered inhabitants as they scrambled for the few available boats. Many died while fleeing the island as boats capsized due to the combined effects of overloading, falling ejectamenta, (ash, and volcanic rock projectiles) and base surge (a lateral explosion of hot gasses that travels at hurricane speed).

The population of the island is 3,628 people in about 600 households. They enjoy a relatively prosperous economy based on fishing, fish-farming, agriculture, mining for scoria, and using their boats to bring tourists to the island. The location of settlements on the island closely relates to particularly rich fertile soils suitable for sweet potatoes and corn. Perhaps as a direct result of the island's attractions, the population is currently growing at the rate of 9.6 per cent per year, more than three times the national average. However, the island only has 215 boats which can accommodate 1,900 people. Thus in the event of a very sudden eruption with minimal warning time, only about half of the population would be able to escape.

Any future eruption similar in intensity to those of 1911 or 1965 would also affect ten settlements that surround the lake and which have a combined population of 76,000 (see Figure 8.4). During a Disaster Management Training Workshop in 1988, arranged by the government of the Philippines, participants visited the island to discuss vulnerability and safety with the residents and with local public officials from the government who lived on the mainland. The group found that there was very little anxiety on the part of the population over the risks they faced, even among those who had survived the 1965 eruption. The lack of escape boats was also of minimal concern, since residents referred to the building that had been set up on the island by the Philippine Institute of Volcanology, and they clearly thought that this was some form of 'volcanic eruption insurance policy', assuming that this agency would be able to look after them in the event of a disaster. They also felt that the very presence of this warning station proved that it was actually safe for them to live on the island. It is doubtful whether the symbolic effect of the structure on the local community entered into official thinking when it was conceived.

Community leaders who were interviewed were much more concerned about the failure of the government to build medical facilities or a school on the island, which resulted in their children having no education or travelling

Box 8.3 continued

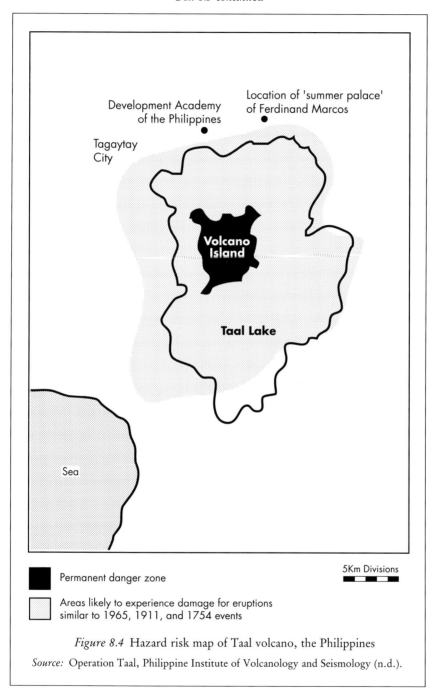

Figure 8.4 Hazard risk map of Taal volcano, the Philippines

Source: Operation Taal, Philippine Institute of Volcanology and Seismology (n.d.).

Box 8.3 continued

each week to the mainland where they lived with relatives during the week while attending local schools.[5]

Government officials responded to the criticism by stating that the island was a designated centre of special scientific and environmental importance. People were not allowed to live on the island on account of the severe volcanic dangers. Therefore, land-use planning controls required the government not to spent any public money on such facilities as schools and dispensaries. Provision of such services would only result in more people coming to live on the island, with a consequent expansion in risk.

The workshop participants, mainly government employees, then responded forcefully that since the residents of Taal Island paid taxes to the government, their acceptance by the government was a tacit admission that the settlement was legal. Consequently, in their view, the government was obliged to provide basic services to the community. They went further and suggested that these services should include disaster-preparedness for the island community which should comprise additional emergency evacuation boats (DSWD n.d.).

The Taal example supports the view that vulnerability to volcanoes is not confined to poor households, or to those who in some other way are marginal. In fact, the Development Academy of the Philippines is situated on the spectacular Tagaytay Ridge overlooking Taal Lake, near some expensive houses owned by wealthy Filippino families. Nearby is also a summer palace that President Marcos half built before his fall from power. This property, including the Development Academy for the training of senior civil servants, is adjacent to the Taal 'high-risk' zone. The records indicate that many people died in precisely this area during the 1911 eruption.

On the other hand, many people in the Taal area have good memories of the Mayon eruption of 1984, when there were enormous economic losses but no casualties in the affected area (which had a population of 90,000) due to a very effective evacuation of 73,400 people (Tayag n.d. (a); Zarco 1985).[6]

The farmers and Marcos took a broad (and private) view, in which Taal volcano was just one of many ingredients (or perceived risks) that influenced their decision on where to build a house. In contrast, the volcanologists, who sought to ban the occupation of the island, took a narrow (and public) view of risk and vulnerability that was confined to the values of their professional background linked with the values of officialdom, and failed to acknowledge that the island was full of useful resources. Each group represents an entirely legitimate and logical response to the same hazard, yet their views vary because their respective needs, priorities, perceptions, and values are very different from each other.

Secondly, and following from the first, local institutions can be strengthened and the capability of families to reduce their own vulnerability can be improved. This is Anderson and Woodrow's (1989) notion of 'rising from the ashes'. However, to achieve this end energy and resources need to be focused on strengthening the self-reliance of the most vulnerable households and their local institutions. We return to the difficult question of identifying and aiding 'the most vulnerable' in Chapter 9.

Dudley assisted local Ecuadorian artisan builders to rebuild their homes in a safe manner after the earthquake of March 1987. He reflected on the experience:

> We have learnt that with outside support, but not external control, and with limited technical objectives the people can achieve great things . . . true development, disaster or no disaster, will only take place through the strengthening of indigenous infrastructures directly accountable to local people.
>
> (Dudley 1988: 120)

Accountability builds trust, and trust allows access to the inner workings of local coping mechanisms. When these are translated into architectural form, there is the possibility of designing low-cost, safer shelter with local people as partners (Maskrey 1989; Aysan and Davis 1992).

Dudley then reinforced the outlook emphasized in this chapter that disaster risk reduction

> is just as much a product of socio-economic factors as technical ones. The best hope for a community's recovery in a disaster is to have a history of strong organization; it is to this end that local institutions must direct their efforts.
>
> (Dudley 1988: 121)

Maskrey (1989) has compiled many such cases where appropriate technology and localized institution building have reduced future vulnerability to geological hazards, especially earthquakes in Latin America.

Thirdly, the disaster provides an opportunity to develop effective risk assessment with good cost–benefit arguments for protective measures. An example is encouragement offered to local authorities by a World Bank team that has been working in La Paz, the capital city of Bolivia, which faces numerous hazards. In a report on the lessons they had gathered from the project the team concluded that risks could be evaluated, quantified, programmed, and addressed with measures that were affordable to the city, even with all its pressing demands on the budget:

> we calculated that disaster prevention and preparedness would cost US$500,000 in 1987 and total about US$2.5 million, or US$2.50 per capita . . . this amount is far exceeded by annual losses from natural disasters estimated at US$8 per capita. With this minimal level of funding, disaster mitigation could be affordable, cost-effective and within the realm of La Paz's needs.
>
> (Plessis-Fraissard 1989: 135)

Finally, disasters provide an opportunity to educate political leaders and decision-makers about the true nature of vulnerability to disaster risk (see Box 8.4). Authorities may be ignorant, or they may deliberately avoid

Box 8.4   Case-study: post-disaster response following the Nevado del Ruiz
volcanic eruption, 13 November 1985, Colombia

A second example concerns the effects of an eruption of Nevado del Ruiz, on
the town of Armero in Colombia on 13 November 1985. Unlike Taal this
volcano had been relatively inactive since its last eruption in 1845.

The eruption occurred at 3.15 p.m. and two hours later the residents of the
town of Armero (population 29,000) noticed that fine dust was falling. At 5.30
p.m., the National Geology and Mining Institute is reported to have advised
that the entire area at risk should be evacuated. By 7.30 p.m. the Red Cross
attempted to carry out such an evacuation, but perhaps on account of very
heavy rainfall from a thunderstorm or the lack of previous evacuation drills,
few agreed to leave their homes. At 9.05 p.m. a strong tremor occurred on the
volcano which was followed by a rain of hot pumice and ash. As a result, part
of the ice cap of the 5,400 metre (17,700 feet) volcano melted and caused the
Guali River to overflow. This in turn caused a natural dam to burst, releasing a
torrent that travelled at speeds of about 70 kms (45 miles) per hour and a
massive mudflow which enveloped the town of Armero (United Nations 1985;
Siegel and Witham 1991).

Within the town one of the few survivors, Rosa Maria Henoa, described
how at about 11.35 p.m.,

'first there were earth tremors, the air suddenly smelt of sulphur, then
there was a horrible rumbling that seemed to come from deep inside the
earth. Then the avalanche rolled into town with a moaning sound like
some kind of monster . . . houses below us started cracking under the
advance of the river of mud'.

(Quoted in UNDRO 1985: 5)

What transpired in those minutes is a horrific story of people attempting to
drive away from the wall of roasting mud, stones, and water. Some died in the
chaos as terrified people tried to climb aboard the moving vehicles (Sigurdson
and Carey 1986; Davis 1988; R.S. Parker 1989; Siegel and Witham 1991).

In the early months of 1988 a group of lawyers inserted a notice in the local
press of the city of Manizales, and in the small towns of Guayabal and Lerida.
These are close to Armero, and housed some of the 3,000 survivors. There was
little point in pinning any notices in Armero since the town had become a vast
deserted 'cemetery' where as many as 22,000 people are buried. The lawyers
invited anyone who had suffered injury, or the loss of relatives or property in
these volcanic mudslides to contact them if they wanted to sue the government
of Colombia for gross negligence in not warning or evacuating them in time to
avoid injury or property losses.

In response to these advertisements no fewer than 750 claims were filed
amounting to a total claim of 20,000 million Colombian pesos (approximately
£40 million sterling). The claim against the government hinges on their alleged
negligence in failing to develop effective preparedness planning (including
evacuation procedures) to enable the population to escape falling debris and
mudslides. These are known hazards that have occurred after previous volcanic
eruptions in the region.

It was anticipated that the government lawyers would argue that the residents were aware of the risks in choosing to occupy a hazardous yet, like Taal

Box 8.4 continued

Island, highly fertile area. By the time the case reached the court in Tolima, the number of claimants had risen to about 1,000. They sued the Ministry of Mines and Energy of the Colombian Government, since the geological service, with scientific responsibility for the issuing of volcanic warnings, was attached to this ministry. The claimants alleged that the government had failed to protect its citizens by not enforcing an evacuation of the citizens of Armero and other affected villages.

The government lawyers argued that the 'ordering of evacuations' was not one of the designated functions of the Ministry of Mines and Energy. But, as proof of governmental concern, they produced evidence that Civil Defence had conducted a 'door-to-door' campaign to warn people to evacuate during the early stages of the eruption. Three expert volcanologists were questioned as to whether the scale, location, and timing of the mudflow could have been accurately forecast. They gave a negative answer, and on this basis the government was cleared of responsibility (Wilches-Chaux 1992b).

The Nevado del Ruiz disaster was a catalyst that had a dramatic impact on the development of disaster protection in Colombia. An effective Governmental Preparedness System at central and provincial level was created, which includes detailed warning and evacuation systems. However, while preparedness planning exists on paper, maintained by governmental legislation, economic priorities can easily dominate safety considerations in practice. For example, the Galeras volcano erupted in January 1993.[7] The nearby town of Pasto is at risk of major eruption from this very active volcano. In 1992 and 1993 the government Disaster Preparedness Agency wanted to issue warnings to the public, but the local authorities refused to authorize them. Their refusal stems from the economic consequences of an earlier warning several years ago, which provoked an immediate financial crisis in the locality when credits and loans were closed (Wilches-Chaux 1993).

recognizing their own role in increasing risks. However, they may respond to messages, such as that of the financial calculation noted above, that action in developing protective measures will be to their benefit later. Quarantelli has emphasized the vulnerability component in disasters and, *a fortiori*, the potential impact of often reasonably low-cost policy initiatives. He notes that

Allowing high density population concentrations in flood plains, having poor or unenforced earthquake building codes for structures, delaying evacuation from volcanic slopes, providing inadequate warnings about tsunamis, for example, are far more important than the disaster agent itself in creating the casualties, property and economic losses, psychological stresses, and disruptions of everyday routines that are the essence of disasters.

(Quarantelli 1990: 18)

In particular, the impact of earthquakes, landslides, and volcanic eruptions will only be reduced when decision-makers become more aware that 'there can never be a natural disaster; at most, there is a conjuncture of certain physical happenings and certain social happenings' (Quarantelli 1990: 18). To conclude, the four verbs that introduced these final suggestions imply the opposite of any passive acceptance of the inevitability of geological disaster losses: to 'change', to 'strengthen', to 'develop', and to 'educate'.

NOTES

1 The earthquake disaster in Peru (31 May 1970) killed 70,000 people, left half a million homeless, and seriously damaged 152 provincial cities and 1,500 villages. Oliver-Smith (1994) argues that the causes must be traced back 500 years to the Spanish conquest and the destruction of the Inca 'prevention mindset' that had created a remarkable ability to mitigate such hazards.

2 The authors, especially Davis, wish to acknowledge the insights and knowledge derived from members of the joint UK–Mexican Government Research Project into Urban Seismic Risk Reduction, 1988–90, especially Yasemin Aysan, Andrew Coburn, Robin Spence, Alexandro Rivas-Vidal, Susanna Rubin, and Hugo Garcia-Perez.

3 Tomblin's point is also evident in another and very rare type of hazard, such as happened on 21 August 1986 in the villages scattered to the north of the volcanic Lake Nyos in Cameroon. A cloud of carbon dioxide gas was emitted from the lake, for reasons that remain unclear. The gas affected everyone living in the area, and asphyxiated 1,700 people, while 5,000 managed to survive the effects (Baxter and Kapila 1989; Sigurdson 1988).

4 Eruptions also produce valuable mineral products such as pumice, perlite, scoria, borax, and sulphur. Residual heat in volcanic regions can also be tapped to provide cheap geothermal energy. The medicinal and recreational use of hot springs has been recognized throughout the world for thousands of years.

5 There is a striking similarity between these complaints and those heard by another one of the authors in a study tour of Bangladeshi settlements on a silt island in the Jamuna River. Residents there did not have access to medical or educational facilities.

6 However, in February 1993 Mayon erupted very unexpectedly, killing forty-seven. About 25,000 people were evacuated from the volcano perimeter.

7 It caused the death of six of the world's leading volcanologists, who were conducting scientific studies in the crater at the time.

Part III

ACTION FOR DISASTER REDUCTION

9

VULNERABILITY, RELIEF, AND RECONSTRUCTION

INTRODUCTION

This chapter develops a number of general principles to guide relief and reconstruction. They are based on the need to ensure that any further increase in vulnerability is avoided, and where possible that it is significantly reduced through the recovery process. Two extended case-studies are presented, the first concerning the major earthquake in Peru in 1970, and the second the famine of 1984 and 1985 in the Sudan. These disasters are chosen because they provide sets of apparently very different problems. Nevertheless, there are important common lessons to be learnt from the recovery process in them both.

The first deals with a predominantly urban population affected by a sudden and overwhelming natural event, and the second with low-density rural populations whose acute shortage of food was triggered by the failure of at least two seasons' rainfall. In addition to these differences, the relief and reconstruction efforts were very dissimilar. After these two accounts there is a discussion of some guiding principles for prevention, relief, and reconstruction that illustrates the affinity between these two very different events.

The Peruvian earthquake, coupled with the 1970 East Pakistan (Bangladesh) cyclone and the three-year Biafra War,[1] which ended in the same year, had a decisive influence on disaster reconstruction worldwide (R.C. Kent 1987a). These were important experiences for governments and agencies, and established patterns of disaster response for the coming decades. They were to be further refined during another massive aid programme set up after the Guatemala earthquake of 1976 (Cuny 1983; Maskrey 1989).

As we shall see shortly, another major episode in forming disaster response policies was the series of African famines (1967–71, 1984–6, and 1988 onwards). During these the relationship between governments and NGOs underwent major changes. It was also during the last of these periods of African crisis, with the combination of civil war and drought in several regions, that an important new international consensus emerged. National sovereignty began to take a secondary place in relation to the survival needs of civilians caught up in war.

Each of the international aid programmes following disasters in East Pakistan, Biafra, and Peru was highly politicized, not least in the US when Nixon was president. During the international relief that followed the Peru disaster the White House took the unprecedented step of appointing their own relief co-ordinator, since they did not regard their own Office of Foreign Disaster Assistance as providing enough 'political image'. The aim was that 'governments will know what we [the US] are doing' (Stephen Tripp, then Coordinator of the Office of Foreign Disaster Assistance; quoted in R.C. Kent 1987a: 77).

THE PERUVIAN EARTHQUAKE OF 1970

This disaster has been selected to introduce this chapter for three reasons. Firstly, this was a very severe earthquake, and its effects were considered to be the worst natural disaster in the history of the Western Hemisphere (Oliver-Smith 1986b, 1994). It is estimated that 70,000 people died and 140,000 were injured (although the category 'injured' is generally difficult to interpret). This earthquake caused more deaths than any other since 1945, apart from the Tangshan (China) disaster of 1976, and was the sixth most severe earthquake of this century. There was in consequence a massive relief and reconstruction operation which provided many important lessons that surfaced in all stages of the recovery process. Secondly, it is possible to look back twenty years with the luxury of hindsight to evaluate the quality of the decisions which were made in managing the recovery process.

Thirdly, a more pragmatic reason for the choice is the existence of a very detailed ethnographic study of one of the affected areas around the town of Yungay. The earthquake induced many rock and mudslides (alluvion), one of which engulfed Yungay. The study covers social change and adaptation in the recovery from the disaster a full sixteen years after the earthquake (Oliver-Smith 1986b). Such longitudinal appraisals written from a social-science perspective are exceedingly rare and valuable.

Oliver-Smith (1994) has also succinctly articulated the most important goal of recovery and rehabilitation, one which eludes many officials involved in recovery planning. He argues that those involved in post-disaster reconstruction need to recognize that we must not recreate 'structures which reflect, sustain and reproduce patterns of inequality, domination, and exploitation'. There is an inherent danger of regarding reconstruction as a strictly physical process, aimed at the restoration of 'normality'. Such a view can fail to recognize that disasters expose the chronic vulnerability of 'normal situations', which, as has been seen in earlier chapters, are almost inevitably a symptom of deep-rooted causes.

Oliver-Smith, an anthropologist with experience of Yungay before the disaster, returned on a number of occasions to describe in vivid detail the plight of the several hundred survivors (pre-earthquake population 5,000). Initially the surviving families were dejected and demoralized by the sheer scale of their tragedy: Yungay, like Armero (Colombia) (see Chapter 8), was totally buried by the alluvion. In most disasters some elements of a town remain but in this case destruction was complete. Oliver-Smith describes how:

the Yungainos were able to weather their individual and collective traumas. They reconstructed their community and lives largely through the choice of secure, known solutions to problems in their new community through a unified social movement motivated largely by adherence to a complex of material solutions and cultural traditions.

(1986b: 261)

Such was the resilience and courage of the community that gradually they began the slow process of recovery.

Such efforts have to be seen in the wider context of the aid efforts that engulfed the country after the earthquake. Kent has commented on the failure of UN agencies in Peru, as well as the 'massive intervention of other donors . . . [who] "invaded" Peru's afflicted northern mountain region'. The result was that vast quantities of aid of all types, 'inappropriate and appropriate, inundated the affected areas, and the Peruvian government's attempts to control the situation were thwarted by the sheer weight of external pressures' (R.C. Kent 1987a: 51). Doughty, an anthropologist who worked in the valley for several months after the disaster noted that

The number of public, international and private agencies at work in the area was truly extraordinary. There are more experts per square inch than fleas on a dog's tail and it would seem that every other vehicle has an official seal of some kind on the door . . . . [T]here appears to be a direct correlation between the velocity at which a car is driven and its official status, as opposed to non-public ownership. Thus, the people are confronted with infinite visits . . . of dignitaries and engineers who hastily make notes, then disappear in the dust. The ones who arrive by car, of course, are less prestigious than those who drop in by helicopter. Those on foot are really in tough shape, but then again there seem to be few of these. The length of time passed in each place also correlated highly with the mode of travel; indeed, one could predict a perfect scale in this respect. It's likely that Galo Plaza [Ex-President of the Organization of American States] established the record, alighting at most places for timed stays of 3.2 minutes.

(Doughty 1970; quoted in Oliver-Smith 1986b: 101)

In the years that followed the departure of the army of visiting aid workers it became clear that there was a heavy price to be paid for all the

197

wasteful, haphazard, politically-motivated assistance. Brian Pratt worked in the region as the Oxfam Field Director from 1978, seven years after the disaster. He and Jo Boyden, a consultant to Oxfam at this time, have described the acute difficulty in the introduction of development projects within the area during the late 1970s and early 1980s, a decade after the earthquake. This was due to the excessive provision of largely inappropriate disaster assistance by foreign governments and agencies, as well as the unfair manner in which much of it was distributed. As a result, local Peruvian development agencies were unable to run effective development projects in the region.

Pratt and Boyden reported (1990) that the post-earthquake legacy was one of local families with a 'hand-out' mentality, still retaining very high expectations that agencies would provide all manner of capital-intensive projects. These were prevalent in the post-earthquake period, when inexperienced relief officials had more cash than they could effectively dispense. This had been donated by generous governments and members of the public. They had responded to the very high casualty figures, and the horrific media images of a town with the vast majority of its residents buried under 6 metres (20 feet) of mud.

Another problem for relief officials was the persuasive influence of companies in the donor countries, anxious for commercial or public relations reasons to be seen to have their products come to the aid of 'helpless victims'. One example from many was the provision of polyurethane igloos, provided as rather improbable disaster shelters to the survivors of the town of Caraz by the Bayer Chemical Company, working in association with the West German Red Cross (Davis 1977a: 35, 1978: 51; UNDRO 1982c: 66–7).

The notions that 'too much aid can be bad', and that vulnerability can persist in subsequent relief operations and even be extended in reconstruction planning, all run counter to popular wisdom. They are not immediately obvious and were not reported strongly enough to counter the myth of the 'phoenix from the ashes'. This stereotypical view colours the way some media staff (or more frequently agency magazine reports) have described the recovery process in glowing terms in which pre-disaster 'wrongs are righted, the hungry fed and the homeless housed'. However, experienced observers such as those quoted in this Peruvian example shared no such illusions.

Sagov (a young architecture student) studied the post-earthquake situation in the affected area. She states:

> the entire valley is littered with half completed apparently abandoned schemes – roads paved only half way, an unfinished experimental fish farm, a derelict brick factory and a school without a roof, to name a few I spotted . . . it is as if the desire to reconstruct mysteriously disappears with the funds.
>
> (Sagov 1981: 182)

She also came to recognize that there was a direct correlation between the earthquake impact and the persistence of highly vulnerable conditions:

> the earthquake only serves to expose and amplify pre-existing problems, i.e. that there isn't enough food; that the agriculturalists of the Sierra (mountains) are migrating in droves to seek their mythical fortunes on the coast; that the corrupt administration, in league with foreign investors, are systematically sequestering and exhausting the country's resources; that there is a permanent housing shortage and a lack of services and amenities – in short a perpetual disaster situation.
>
> (Sagov 1981: 179)

Pre-disaster inequalities were replicated in the reconstruction.

In a pattern familiar in many similar situations, the provision of post-disaster housing went first to the landowners, second priority were land-owning disaster victims. The impoverished peasant families who migrated to the area from the surrounding region after the earthquake were at the end of the queue. As a result:

> a city has been built which in its very design reinforces class divisions and undermines the formation of any community solidarity. The delivery of housing aggravated class relations and, in the final analysis, the housing itself has accentuated the conflict and became one of the clearest markers of class differences in material, spatial and symbolic terms in the new city.
>
> (Oliver-Smith 1990: 16–17)

The pattern of decision-making for the construction of new housing in Peru was largely determined by central government. A firm of architects from Lima were awarded the contract to redesign the town of Yungay.

A fierce conflict developed between the government's reconstruction agency and local residents over the relocation of the regional capital. Ultimately, the residents won, and the battle forged a new sense of purpose and solidarity in the community (Oliver-Smith 1986b: 51). The Yungay survivors also insisted that they resettle at the site of the old town. When the government objected, the survivors threatened to camp out in the ruins and alluvion in full view of international news cameras. Again, the survivors' wishes were finally honoured.

During the long and painful process of reconstruction it became clear that there were tensions between the desire for change and the urge to restore what had been. For all its limitations and dangers, the pre-disaster environment had established living patterns for the survivors, provided the framework for their livelihoods, and was a physical expression of their culture:

> ultimately, engineers, architects, planners and social scientists, in assuming their responsibilities in reconstruction, are faced with

199

walking a very fine line between a stricken population's need for continuity and the design of a community which will sustain the further development of social institutions of greater equity and justice.

(Oliver-Smith 1990: 17–18)

To this we might add greater safety.

SUDAN 1983–92

The famine of 1983 to 1985 claimed the lives of perhaps 100,000 people in the Sudan, mainly in the Darfur, Kordofan, and Red Sea regions. Mortality was very variable spatially and across groups, though generally the more arid north suffered the greatest proportion of deaths. This was particularly so in areas which relied heavily on livelihoods that depended on rainfall, especially growing millet and sorghum and rearing livestock.

The groups which suffered most were displaced people who migrated to camps and then returned home, pastoralists with small herds, and sedentary agriculturalists lacking substantial food stores. Those who had access to incomes not so reliant on rain (e.g. weaving, woodcutting, or urban employment) were least affected (de Waal and Amin 1986). Livestock also died in very large numbers (sometimes as high as 70 to 80 per cent in northern Darfur).

Where does a chronicle of famine start? The point at which it emphatically does not start is the trigger provided by the occurrence of a natural event, in this case a drought. A number of much longer-term root causes and pressures were to blame (see Chapter 4). Two were fundamental and these were the decline of the ability of individuals, households, and families to provide for themselves, and the reduced capability and willingness of the Sudanese state to prevent famine and to provide timely relief when serious food shortages were clearly forecast.

The first root cause involves a number of complex and interrelated factors analysed by de Waal (1989b). The overwhelming and obvious drought in the Sudan was an immediate trigger to famine. The drought had been unremitting for more than a decade. Comparison of the isohyets (rainfall 'contours') of 1976 with those of 1985 shows that the pattern of rainfall moved southwards by about 100 kilometres (60 miles).

The ability of some groups to maintain their livelihoods and food sources also seems to have been reduced because of environmental degradation in the region. Quite how this may relate to a long-term decline in rainfall is uncertain and controversial. Crop yields had been in decline, and some consider this to be partly a result of environmental damage. But there are of course many other possible causes of this, including reduced rainfall.[2]

The crucial issue is to understand the characteristics of the different groups of people and their vulnerabilities in the face of a combination of factors that changed their livelihood opportunities at the same time. Of these factors, reduced rainfall was only one. While it would be foolish to consider that the people of this region could survive any number of years of drought without some impact on their well-being, they are accustomed to severe fluctuations and variations in rainfall. But other factors affected their ability to mitigate the drought's impact, prevented the use of some conventional coping mechanisms, and increased the number of people and livestock competing for the same limited resources.

Local people in Darfur understand well how drought and failure of rain-fed crops have forced people to cut down trees for firewood as a last resort income-earning opportunity (de Waal 1989b). Also classic symptoms of soil erosion are not hard to find and are recognized as 'violations of the rural landscape' or 'things in their wrong places' by local people (de Waal 1989b).

Such localized but probably significant deterioration of both range and rain-fed arable land has undoubtedly been brought about by unprecedented concentrations of people, which have built up as a result both of refugees from war in Chad and from desertification to the north. Herders have lost stock to continuing droughts and have become sedentary, competing for land with existing agriculturalists. Fallowing periods have been reduced and cultivation extended on to more fragile soils. Natural population increase in humans and livestock has also been rapid. These interlocking factors – drought, some environmental degradation, increasing concentrations of people competing for limited land, and declining yields of livestock and grain – have all served to undermine people's ability to self-provision.

Darfur and other districts in western Sudan remained relatively isolated from the national grain markets. However, the west had been integrated into other cash-crop markets, principally groundnut and sesame, and the much longer established gum arabic trade. These crops did provide additional incomes for some farmers during the 1960s and 1970s. But in the 1980s drought, a breakdown in railway communications (due to a struggle between the government and rail unions) forced the region to retreat from the market with a consequent drop in incomes for some groups.

So far, this account attributes the famine that struck the people of western Sudan to long-term food availability decline, accentuated and brought to acute levels by repeated failure of the rains. It is undeniable that these factors form a crucial part of the explanation. But they have to be seen in the context of the other factors that disrupted the people's capacity to deal with such a situation. Moreover, a failure to be self-sufficient in food is neither a sufficient nor a necessary cause of famine. The explanation also has to look at the role of the state, and at more detailed issues of people's access to different occupations, their location, ethnicity, and gender.

After independence, the state became less capable and willing to maintain the famine prevention policies that had operated reasonably well in the colonial years.[3] Then from 1978 mismanagement, corruption, and economic crisis on a massive scale became the norm. Foreign debt escalated alarmingly, reaching US$9 billion by 1985, and exports fell to almost zero. Corruption itself came to be recognized as the 'fifth factor of production' (Kameir and Karsany 1985, discussed in de Waal 1989b). During 1983–5 food availability in many parts of the Sudan became an acute problem. Neither the presence of many thousands of migrants in western Sudan and in camps near Khartoum, nor pressure from the regional government in Darfur were enough to persuade the central government to prepare for the coming crisis.

Even after the disastrous rains of 1984 the government failed to declare a national famine, for similar reasons of national prestige as the Ethiopian government at a similar time. It was possible to convert production on state-controlled irrigated areas from non-food crops to the production of sorghum. However, the transfer of grain from the surplus-producing eastern region was prevented by merchant interests. Also, Shepherd claims (1988: 61) that the government was anxious to maintain the export of sorghum to help Sudan's deteriorating balance of trade position after 1978.[4]

The economic interests of merchants and landowners were promoted by the state, against the interests of those vulnerable to famine. The distribution of grain to the west was hampered by inadequate labour provision and administrative incapacity of the Food Aid National Administration (FANA). Also, the railways had been run down, staff were on strike during crucial periods of the food shortages, and the contract for transporting food aid was awarded to a greedy private contractor. Roads had declined to such a degree that the journey from Khartoum to the western regions could take over five days. International aid agencies were alerted too late and most of them responded far too slowly.

Within western Sudan distinct patterns of hunger emerged. One attempt to map people's survival time (termed famine vulnerability in the study) was attempted by Oxfam and UNICEF (modified by Ibrahim 1991: 191–2). Figure 9.1 shows the distribution and degree of survival time (vulnerability) in Darfur in March–April 1986. The method involved allocating points (on the basis of village council areas, not individual or household) indicating survival time without extra food, with higher points indicating low survival times, high levels of malnutrition, lack of stocks of food or animals, and high prices for millet. The map shows a belt of chronic food shortage running northwest to southeast through El Fasher. This was an area of exceptionally high population density due to the influx of refugees from the desiccated north.

Clearly there are other factors involved, particularly whether people had access to other income opportunities not directly dependent on rainfall. The vast majority of people, though, were reliant on incomes derived from rain-fed agriculture and livestock. Since the drought was so absolute for 1983–4, grain production was negligible in north Darfur. Production further south on better soils and with higher rainfall was able to satisfy 90 per cent of the

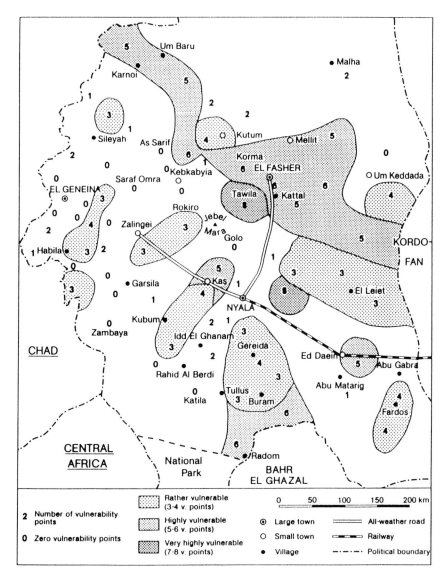

Figure 9.1 Distribution and degree of vulnerability in Darfur in the aftermath of the famine disaster 1985–6

Source: Ibrahim (1991: 193), after Oxfam.

needs of the population in 1983 but less than 60 per cent in 1984. Those who had been able to store grain from 1982 and from any harvest there was in 1983 and 1984, clearly suffered least. There was a general move by both farmers and pastoralists to low status income-earning opportunities (agricultural wage-labour, charcoal-burning, firewood collection, and pottery).

These trades had to be plied near clients who predominantly lived in towns. Many destitute people had moved to regional and smaller towns and had taken up these trades. However, in the classic manner of progressive entitlement failures, disposable incomes for the purchase of these goods and services also dried up (Cater 1986). Those fortunate enough to be in government employment had to riot in order to persuade the government to provide subsidized sorghum. Other disbursements of grain, when it finally arrived in July 1985, mainly went to those who had settled near towns, and hardly any found its way to the pastoralists.

Another aspect of access to resources was the ability to mobilize community support systems and (ancient) coping mechanisms (see Chapter 3). Various means of transferring resources in times of food shortage existed, including 'Zakat' and 'Karama' (religious feasts) and 'food-for-work' programmes patronized by richer members of the community. In general, the displaced suffered more, since they lost contact with community support networks.

It should be emphasized, however, that the degree of support from the community was very limited and later on during the famine collapsed altogether, since the numbers able to give it were minimal in comparison to the numbers in need (Pyle and Gabbar 1990). In addition there were famine foods such as 'mukheit' (*Boscia Senegaliensis*) and 'koreib' (*Brachiaria laetun*), but these were particularly scarce near concentrations of population.

Thus the access position of different people to the productive resources of agriculture and livestock, to low status income opportunities (although access here is tempered with willingness to enter such occupations), to community coping mechanisms, and to famine relief supplies determined people's changing vulnerability to eighteen months of hunger.

PRINCIPLES FOR MANAGING DISASTER RECOVERY

The following principles can be suggested from the various observations on the Peru earthquake and Sudan famines as well as the material presented in Part II of this book.

Principle 1: Recognize and integrate the coping mechanisms of disaster survivors and local agencies

The primary resources in the relief process are the grass-roots motivation and collective efforts of survivors, their friends, and families. Other groups can help, but they must avoid doing anything best undertaken by the survivors themselves (UNDRO 1982c).

In Kenya, nearly twenty years ago, one of the authors documented a wide range of more than seventy coping mechanisms that rural people used to survive drought. These ranged from boarding small children away in the home of a more fortunate member of the extended family to reliance on non farm income and use of wild famine foods (Wisner and Mbithi 1974; Wisner 1978b). Returning to those villages in 1990, he found many of these mechanisms still in place. But in addition, coping now included neighbourhood-based women's self-help activity, and highly developed knowledge of how to 'play' the aid and relief system. Others have commented on the richness and diversity of ordinary people's coping strategies, and we have reviewed this literature in Chapter 3.[5] It is important to remember that local coping knowledge, especially in process of recovery, often varies according to gender (Indra and Buchignani 1992), age (Guillette 1992), and class (Winchester 1986, 1992).

In summarizing how local 'coping mechanisms' relate to vulnerability in situations as diverse as the ex-USSR and Sub-Saharan Africa, it is worth noting the attitude of relief officials in the Armenia earthquake disaster of 1988:

> ad-hoc search and rescue groups worked hard for days and nights, but they had no hand tools such as picks, shovels, and axes to help them in their efforts. Moreover, since most officials concerned with the formal bureaucratic emergency response were focused on getting bureaucracies to the site, no one thought to send hand tools to the victims; victims were not envisioned as part of the solution in search and rescue; instead, they were viewed as part of the problem – people who needed food, water, housing and so on.
>
> (Armenia Earthquake Reconnaissance Report 1989: 153)

In fact most of those dug from the rubble within the first few days were rescued by survivors, not the official rescuers. The response of survivors offers a basis for further recovery activities, but often its potential is not recognized.

The imperative of building on existing coping behaviour extends one step further to reinforcing the efforts of local organizations. In Bangladesh following the 1991 cyclone, some 100 foreign and seventy Bangladeshi NGOs began to provide relief. But in general they failed to take advantage of the existence of about twenty NGOs that were already working in the affected area (Sattaur 1991: 23). This same pattern of 'swamping' by outside agencies occurred in the case of Peru, discussed above, more than twenty years earlier. Nor is this problem unique to the Third World. Following the Loma Prieta earthquake in California in 1989, the Red Cross and Federal

Emergency Management (FEMA) officials were reluctant to include local organizations in post-disaster planning even though some, such as an Hispanic medical clinic in Watsonville and the Farm Workers' Union, were engaged in relief and recovery activities (Laird 1992).

Principle 2: Avoid arbitrary relief assistance

The scale of relief assistance is not normally determined by any accurate assessment of needs and damage, but rather by such arbitrary factors as numbers killed, political constraints, the time of the year, the accessibility of the affected area, the extent of national and international media exposure, and the consequent volume of cash from donors.

The arbitrariness of international relief is clear in the situation of Nicaragua, where, just six weeks before the Armenian earthquake in 1988, a severe hurricane hit the Atlantic coastline (see Chapter 7). Normally if a disaster of this destructive power had occurred in the Caribbean region, such as hurricanes Gilbert and Hugo not long before, there would have been extensive international aid. But the political tension between the US and Nicaragua at the time suggests that the US used its influence among its allies to enforce a virtual aid embargo. Perhaps this would have changed if there had been very high casualty statistics, when the public might have exerted pressure on their governments to give assistance.

Not only was aid denied, there was no recognition in the western press or academic journals of the remarkable evacuation of the population before the hurricane struck the coast. Within just 72 hours the Nicaraguan army succeeded in evacuating over 300,000 people from the exposed coastal regions, probably saving hundreds of lives in the process.

Another form of arbitrariness in relief to be avoided is inappropriateness of material aid, regardless of the source. Given the considerable experience of relief built up in recent years, there can be no excuse for the provision of culturally unacceptable foodstuffs, spoiled or tainted food, or shelter designs that increase vulnerability in the long run. Yet inappropriate aid has continued to flow unabated. In Bangladesh following the 1991 cyclone, some NGOs began using corrugated aluminium roofing. Local inhabitants were concerned that these could become deadly flying objects in high winds, and that traditional thatched roofs serve a double function as rafts that can save the drowning (Sattaur 1991: 23).

This is not to imply that 'appropriate' material relief has to be 'traditional', local, or 'low-tech'. For instance, modern water sanitation systems are vital in many refugee or displaced person camps. In the Bangladesh case just mentioned, 50,000 packets of vegetable seeds were imported by the Mennonite Central Committee to enable farmers to grow a quick crop while work continued to rebuild embankments that protect arable land from the sea (Sattaur 1991: 22). This was probably justified, although the seed was an exotic import arriving by air freight.

'Arbitrariness' is not just a matter of hastiness or ignorance on the part of donors, but concerns who defines the terms of reference, who determines what 'needs' are in the first place (Wisner 1988b; Oliver-Smith 1992). If the entire relationship between donors and governments and ordinary people during 'normal' times has been characterized by outsiders deciding what 'development' means and what 'needs' are, then it is not surprising that during a crisis this external control persists.

The supply of relief and reconstruction assistance from external groups is strongly influenced by a number of arbitrary factors such as geopolitics, media coverage, and commercial interests. So it is essential as far as possible to rely on local and national resources, developed within a well-organized preparedness plan.

Principle 3: Beware commercial exploitation

National and international commercial firms, often working in convenient partnerships with relief agencies, are extremely active in reconstruction situations. Such companies can exert pressures on local communities or even governments which may result in wasteful, culturally inappropriate solutions that fail to regenerate the local damaged economy.

The theme slogan for the Prefabricated Building Manufacturers Association Exhibition after the 1980 south Italian earthquake was 'One Man's Disaster is Another Man's Marketing Opportunity' (noted by Davis). Indeed, following that earthquake many manufacturers in northern Italy reaped great profits, especially those producing prefabricated housing (Chairetakis 1992). In other disasters, overseas manufacturers have convinced donor agencies or government officials in one way or another to purchase items that turned out to be inappropriate, faulty, or substandard.[6]

Principle 4: Avoid relief dependency

Relief assistance, if poorly managed, can create unrealistic expectations and long-term dependency, and therefore can be a serious obstacle to subsequent developmental assistance.

A major reason for this is the excessive aid that normally goes into relief assistance, while reconstruction, preparedness, and mitigation are often starved of the necessary resources. Disaster assistance needs to be provided to the affected community in a manner which enhances community self-reliance. It must do this by enabling people to make the key decisions about their future welfare, and to become actively involved in every stage of their own recovery, without any paternalistic, dominating pressure or inter-ference from assisting groups.

Too often survivors are relegated to the role of passive spectators by aid workers who rapidly take over the entire recovery process. This destructive tendency has been noted in a wide range of situations including the after-math of many of the earthquakes, storms, and floods discussed in Part II, as well as in longer-term displacement of refugees by famine or war (Harrell-Bond 1986).

Every opportunity should be taken to encourage the affected population to serve their own needs. Health workers, teachers, and craftspeople should be able to use their skills and, indeed, such human resources have been successfully integrated into relief and recovery efforts (Anderson and Woodrow 1989; Maskrey 1989). In the same spirit, food-for-work pro-grammes can be less demeaning than prolonged hand-outs of relief food, and provide a sense that the shattered world is being physically reconstructed. Where income is also provided for such work, there is an additional benefit, because the people can use that money to rebuild and normalize their lives (see Chapter 4). Donated wheat was used in Bangladesh in 1991 to recruit many thousands of workers to reconstruct embankments to protect crops from further sea flooding.

However, care must be taken to avoid abuse. In the early 1980s food-for-work was used during a drought in northeast Brazil, but was criticized because the water storage works being constructed would only benefit large landowners. Food-for-work appeared to be a windfall for the rich. In Lesotho at about the same time one of the authors visited a village water tank painstakingly excavated by hand by women receiving food-for-work. What took the 150 women eighteen months would have required a bulldozer for only a few days, and the women knew and resented the fact. The appearance and reality of 'made work' should also be avoided.

Principle 5: Decentralize decision-making when possible

After a major disaster, decision-making tends to be a centralized process, possibly due to media pressure or the inevitable high political profile of such events. So local officials, who may be accustomed to making local plans in conjunction with their citizens, could easily be bypassed and the consequences could be negative.

This is not simply an issue of getting the balance right between central plans and local participation. It is about the process of empowerment, appropri-ateness of assistance, and recognition of people's own capacities:

a distinction must be drawn between the necessity for centralised preparedness planning and the equal necessity for decentralised implementation at the local level. Sectoral and ministerial conflicts threaten effective policies. If decision-making at local levels is taken to central government, there is a weakening of the authority of local officials . . . centralisation also results in both human and material assistance flooding the area from central sources, thus reducing the opportunities for the revival of the local economy, or for local officials to take responsibility for their own structures.

(Aysan and Oliver 1987: 19–20)

Principle 6: Recognize disasters as political events

Major disasters are inevitably important political events at local, national, and international levels. Therefore they are frequently exploited for short-term political gains which may conflict with assistance on the grounds of humanitarian and developmental criteria.

In some cases, the ensuing political struggle may result in a more, rather than less, equitable distribution of aid and even longer-term political benefits (and reduced vulnerability) for formerly less powerful groups. For instance, after the Loma Prieta earthquake of 1989, Hispanics won city council seats and elected a mayor in Watsonville, California (Laird 1992).

Most experienced disaster relief workers have encountered the tendency for local elites to try to capture relief for their own ends:

> political considerations often take precedence over humanitarian or sound disaster management considerations. The leadership of a country may recognize disaster response as an opportunity to curry favour with a constituency by offering preferential treatment. Politicians may divert disaster relief to other uses: simple corruption of diverting material assistance to the pockets of people in power or, in the worst cases, to support military operations which repress disaster victims.

(UNDP 1990b)

Principle 7: Recognize pre-disaster constraints

Reconstruction activity is constrained by pre-disaster limitations and deficiencies.

These range from structural weaknesses in the national economy to political and administrative constraints. The building industry may be dependent on imported items which the country cannot afford, especially if the disaster has cost it foreign exchange because of disrupted export trade. Such

209

disruption is common: destruction of roads prevented coffee exports from Guatemala following the 1976 earthquake; a ruptured pipeline cut off Ecuador's export of oil for many months following the earthquake in 1987.

Government budgets may be overextended due to civil war or destabilization, as we saw in Part II when discussing recovery following coastal storms and floods in Mozambique. In recovering from drought, a number of countries of the Sahel failed to encourage smallholder food production because of their commitment to national economies built on exports (beef, groundnuts, cotton, coffee, tea) and minerals (oil, uranium, phosphate).

Principle 8: Balance reform and conservation

In reconstruction planning there is always the need for reform to introduce mitigation measures as well as social, political, and economic changes to reduce inequity and vulnerability. But there is also a parallel need for continuity with the past.

Any physical or social reforms have to be finely balanced with conserving familiar elements of the pre-disaster society and its setting. As Koenigsberger has observed:

> the immediate post-disaster period provides the planner (reformer) with two unique advantages: (a) Survivors are ready to accept change; (b) The public is ready to provide funds . . . . On the other hand there are strong forces against change: (a) Frightened people and, even more, frightened authorities are wary of change. Reform needs courage. (b) Pre-disaster conditions, however bad, appear to victims in a rosy light.
> (Quoted in Davis 1978: 66)

Oliver-Smith (1986b) found a similar interplay between reform and nostalgia in Peru during reconstruction following the 1970 earthquake.

Principle 9: Avoid rebuilding injustice

Relief and reconstruction can aggravate divisions and patterns of inequity within a society. Social, economic, and political vulnerability are often reconstructed after a disaster, thus reproducing the conditions for a repeat disaster.

Case-studies from Part II of this book contained many examples of what should be avoided. Due to unequal access to resources in rebuilding, the gap between rich and poor farmers in the Krishna delta of India was greater eight years after the cyclone than before. In Bangladesh, rich landowners are more likely to gain access to new farming land created by floods. Kenya, Nigeria, Sudan, and other African countries provide cases where drought is an opportunity for the rich to acquire more land and livestock at the expense of the desperate smallholder. After the 1976 earthquake in Guatemala, official recovery widened the gap between rich and poor, and the resulting resentment ushered in a period of bloody repression by the army and death squads (1978–82).

Case material has also revealed how official relief failed to address the marginality of poor rural Blacks in North Carolina or Hispanic farm and cannery workers in California following hurricane and earthquake, respectively (Miller and Simile 1992; Laird 1992; Johnston and Schulte 1992). After the 1980 earthquake in Campania (southern Italy) most of the US$1.9 billion in relief from the European Community and eighteen other countries was spent with Italian contractors and industries from the north, as was the US$3 billion the government provided for rehousing. The Italian government's programme for industrialization in the affected mountain valleys used up scarce agricultural land and polluted the rivers with factory effluent. Unique mountain village cultures and ecologies were destroyed, further weakening potential local coping in the face of future disasters. On balance, the pre-existing inequalities between north and south in Italy were strengthened (Chairetakis 1992).

Principle 10: Accountability – the key issue

The key to success or failure of recovery is the extent to which the assisting groups make themselves accountable to the recipients for the services they provide.

Krimgold emphasized this element more than fifteen years ago:

> disaster relief efforts must fundamentally be directed to meeting the expressed needs of the victim populations. They must be enfranchised and given choice in the relief process. Further, any collective mechanism for the improvement of relief activities must be based on the evaluation and critique by the victim population. Mechanisms must be developed which tie donor agencies to the judgement of the supposed object of their activity.
>
> (Quoted in Davis 1977b: 22)

We saw in Part II that survivors often have to force governments to listen to their opinions and desires. Accountability is seldom provided freely by bureaucrats who are used to functioning in a distanced, hierarchical manner during 'normal' times: it is terrain of struggle for power and control.

Principle 11: Relocation is the worst option

Plans to relocate entire settlements after a disaster on grounds of hazard mitigation are rarely desirable. Such operations are very costly, and provoke social upheaval and discord precisely when the survivors need a stable environment to aid their recovery.

Normally such plans stem from political sources or landowning interests, and use the hazard argument only as a pretext. In Part II we saw a number of cases of resistance to such relocation, including Mexico City following the 1985 earthquake, in Ethiopia during the famine of the mid-1980s, in the Rufiji valley (Mozambique) after floods in 1969, and with the 1970 destruction of Yungay, Peru. There is also considerable evidence of the serious health and social consequences of large-scale resettlement (Hansen and Oliver-Smith 1982; Harrell-Bond 1986).

The relocation of tens of thousands of people from the north of Ethiopia to the southwest during the famine of 1984–5 was motivated by the government's desire to undermine support for opposition groups fighting in Tigray Province. The resettlement resulted in many thousands of additional deaths due to disease in the reception zones (Clay, Steingraber, and Niggli 1988; Kebbede 1992).

Principle 12: Maximize the transition from relief to development

Relief creates dependency, so it is vital that as soon as the emergency needs are satisfied there is a return to a development approach.

This principle takes us from immediate relief and post-disaster recovery and into the next chapter, where it is argued that mitigation – the creation of a safer environment – must be part of 'business as usual'. Likewise relief and recovery activities should overlap naturally and gradually with 'normal' development activities of governments and non-governmental organizations.

This principle assumes that there had been 'normal' development activities before the disaster (e.g. large-scale industrial or infrastructure projects, primary health care, education and training, agricultural and resource management extension). This assumption might be false in two ways. Firstly, there may have been very little happening in the affected area before the disaster. Budget cuts during the 1980s and 1990s reduced such activities in many parts of Africa and Latin America. Secondly, such activities may well have contributed to the disaster. This could be true where government presence was primarily focused on mega-projects such as mining, hydroelectric, forestry, or irrigation development whose full social and environmental effects had not been analysed from the point of view of disaster vulnerability (see Chapter 10).

In the case-study of Peru, we saw that relief was delivered in such a manner that grass-roots development projects of a self-help variety were not possible for many years following. This was because the character of the aid created dependency and unrealistic expectations of material aid in the affected areas. People who suffer disasters generally want to re-establish their lives as quickly as possible. They do not want hand-outs, but grants or loans to rebuild livelihoods. 'We want work, not relief' was the comment reported from Bangladesh following the 1991 cyclone (Sattaur 1991: 21).

The way recovery is assisted should provide the necessities for healthy and sustainable livelihoods. Too often rural people are abandoned when the official books on a disaster are closed. The opposite approach can also be dangerous, for instance, to equip displaced families with farm implements, buckets, seeds, etc., and expect this to be all that is necessary for them to succeed. The relief–development transition requires more careful preparation. For instance, UNICEF provided drought sufferers in Mozambique with oxen for ploughing and seed from neighbouring Zimbabwe. Both were familiar and suitable.

In a very different social, economic, and geographical setting, 'development' work continued among the rural poor on the islands off Charleston, North Carolina, long after official 'recovery' from hurricane Hugo ended. Churches, in particular, had been sensitized to the existence of an isolated, economically and politically marginal population on these islands. The people's housing needs were great even before the hurricane, and official 'recovery' did not solve their problems. Federal relief money would pay for a new roof, but not for replacement of the pre-existing rotten wall on which it had to sit (Miller and Simile 1992).

PUTTING IT ALL TOGETHER: DISCUSSION OF PRINCIPLES

There is a strong logic that unites these twelve principles. Our suggestions for principles to guide relief policy are empirically derived and point towards common social and political preconditions. Accountability and political will are the most important points connecting all the principles. These are imperatives in donor countries as well as in the affected nation.

Educating about the global–local connection

The learning about relief and reconstruction of recent decades has led to a new spirit of humanitarian internationalism, manifested by Operation Lifeline Sudan and other operations in, for example, Iraqi Kurdistan and Somalia. National sovereignty is no longer assumed to be supreme in all situations.

This shift in thinking has a parallel in the outreach and public education activities of many relief and development organizations in the industrialized countries. Pioneering efforts were made by organizations like War on Want in Britain, Oxfam America, Food First, and smaller groups such as

Grassroots International (US) and organizations such as IRED in Geneva, Aktionsstelle Dritte Welt in Berlin, and Euro-Action Accord. The message was that international economic justice and local empowerment were central issues.

Some northern NGOs today work on poverty, oppression of minorities, and vulnerability to disasters simultaneously in their home countries and overseas. The evolution from nineteenth-century notions of 'charity', through mid-twentieth-century 'modernization' of 'them by us', and uncomfortable late twentieth-century 'partnership' has begun to give way to genuine 'solidarity'. A Zairean development worker finds a place working with youth in Manchester, New Hampshire. Burkinabe goat farmers provide help to their French counterparts trying to cut veterinary costs. A Chilean agronomist provides leadership for sustainable agriculture in California. Zairean prostitutes become resource persons for international efforts in AIDS education (Schoepf 1992).

Such changes require new outreach and public education messages by the northern agencies. There is some evidence that this is happening, and the result is a group of citizens more aware of the root causes of disasters and the goals of relief and recovery. Public pressure on northern governments and corporations can thus be applied, limiting the problems identified in the list of principles above: arbitrary relief and commercial exploitation of relief, for instance.

Learning from local people

Southern NGOs have also been very active in the years since the disasters in Peru, East Pakistan, and Biafra. The integration of indigenous coping into relief and recovery is never easy, especially in hierarchical systems where there is a long history of antagonism between town and country, peasant and landlord, high culture and low culture. Integration of people's knowledge is not a mechanical activity, but begins with respect for the people concerned and requires their trust. Southern NGOs have led the way in demonstrating respect and building trust.

The fruit of these efforts has been the emergence of a new development framework based on people's knowledge and local organization. This approach requires a genuine listening to local people (Pradervand 1989) and an awareness of how power relations can block the participation of the most vulnerable. Indeed, as Chambers (1983) puts it, one must 'put the last first'. Doing so opens up a channel of communication between the people and disaster aid workers that goes beyond 'consultation'. People are able to express their needs and work together with outsiders to overcome obstacles (Lisk 1985; Wisner 1988b). In a more specialized domain this participatory method of working has been called the 'farmer first' approach (Chambers, Pacey, and Thrupp 1989). Water projects, sanitation work, reforestation, housing, grain storage design, and many other efforts have benefited from participatory or 'action research' methods, in which outsiders and local people are equal learners and teachers (Wisner, Stea, and Kruks 1991). Successes have been registered in Asia and Latin America (Conroy and Litvinoff 1988; Holloway 1989). In Africa this approach to 'development with' the people has been called 'the silent revolution' (Cheru 1989; cf. Rau 1991; Kiriro and Juma 1989).

One example points out the potential of popular control of the recovery process. Following a landslide, residents of the Caracas (Venezuela) neighbourhood of Nazareno refused to move into remote barracks provided by the government. Instead, they broke into and occupied a local school. They demanded land and assistance in building more secure homes. The government agreed, and a team including architects and social psychologists from the Escuela Popular de Arquitectura (People's School of Architecture) worked with the community as resource people (Wisner, Stea, and Kruks 1991: 282; Sanchez, Cronick, and Wiesenveld 1988).

CONCLUSION

In considering why problems occur so frequently in relief and reconstruction, it is important to note that as well as residual, pre-disaster deficits, many of the 'actors' involved in the long-term recovery process tend to be decidedly short-term, transient beings. This applies particularly to some international consultants and the staff of certain agencies that tend to sprout as quickly as mushrooms after disasters that attract media coverage. The sudden arrival of newcomers to a disaster is not confined to foreigners, and can equally apply to national government officials involved in relief and reconstruction, or to architects, planners, and engineers involved in rebuilding activity. In the case of remote rural disasters such as the Peruvian earthquake, it is probable that many officials ventured rather reluctantly from Lima, the capital. Such officials typically do their job, often under pressure, but nevertheless with sublime confidence and then all too quickly depart from the scene for yet another disaster or administrative task or commission.

The result of such isolation or lack of public accountability can easily constitute a second disaster, where mistakes can adversely affect the thousands of disaster survivors who deserve better from their government, agencies, and professional groups. Thus the administrators and professionals who provide well-meaning advice or make decisions that create subsequent problems for the surviving community are likely to remain ignorant of the impact of their involvement.

Without adequate mechanisms to inform officials or their decision-making directors of the long-term outcome of their actions, there is minimal learning from failure or even success. Thus the wheel can be reinvented after any disaster, and cursory examination of disaster reports indicates the repetition of failures, most of which could have been readily anticipated.

This raises the question whether we are correct in regarding the experience gained from the Peruvian earthquake as being the 'formative experience' for agencies and governments we noted earlier. While some of the matters noted in Peru have recurred all too frequently, nevertheless some very important lessons have been learned. Now, more than twenty years later, as we reflect on the way relief and reconstruction were managed in Peru or on other events in 1970, it is possible to observe signs of genuine progress.

There is a growing recognition on the part of disaster relief workers that reconstruction planning has to begin by rebuilding many other items than broken walls, roads, or water-pipes. Confidence and trust may need to be established. Institutions need to be formed. Industries, such as the building industry (which may have manifestly failed), may need major rehabilitation, and inequitable patterns of landownership may need to be changed. There is growing acceptance that many of these 'enabling structures' can be 'built' in parallel with visible reconstruction, and it is vital to take action while there is the political will to commit resources.

Even where there is no political will on the part of the elite, the unity of the people often forged in the post-disaster situation can be used to demand reforms. Local mass organizations (trade unions, farmers' unions, women's movements, consumers' unions, associations of professionals such as doctors, nurses, architects, lawyers) and local non-governmental organizations (church-based and secular) can be key agents in demanding accountability and a voice for the people in the recovery process. They are also important in lobbying for social and economic changes that reduce vulnerability in the long run. We take up that subject in the next chapter.

NOTES

1 Biafra was the name adopted by southeastern Nigeria, home of the Igbo people, in an attempt to secede from the Nigerian Federation. During the civil war Biafra was blockaded by Nigeria, and perhaps a million people died of hunger. See Jacobs (1987).
2 Associated with the ideas of climate shifts are pessimistic accounts of 'desertification' where the desert margin advances so many kilometres a year (e.g. Lamprey 1976). But these have to be set against the critics (e.g. Bie 1990; Nelson 1988). However, it is likely that a combination of drought and resulting reduction in vegetation cover together with rapid devegetation (World Bank 1989) have contributed to increased wind erosion on the more fragile *qoz* soils as well as aluminium saturation on other soil types, among other problems (Huntings 1986; MASDAR 1987).
3 These were similar to those developed in India with the Famine Codes, described in Chapter 4.
4 Similar exports of sorghum to the European Community for use as cattle feed in 1992 have provoked much concern given the hunger problems that remain in much of the country.

5 People's ingenuity in coping with hazards is not limited to the rural Third World. Following the Exxon Valdez oil spill off Alaska, citizens of the coastal zone and islands invented a series of new technologies such as beach cleaning machinery (Button 1992) and oil containment booms made from 'geotextile' woven with established net-mending techniques (Mason 1992).

6 One of the authors was a lunch guest of the British ambassador to Nicaragua during a visit three months after the 1972 earthquake that devastated the capital Managua. The ambassador's aim was to encourage him to promote a British contractor which produced prefabricated hospital buildings when he met Nicaraguan officials. It was suggested that he should encourage the Nicaraguan Reconstruction Committee to place an immediate order for four British hospitals to replace those destroyed in the quake. Asked by the author (who is an architect) whether they were earthquake impact-resistant, the representative commented 'God knows'. Having indicated that he would not therefore advocate such a product, the author was advised by the ambassador that it was their job to promote British trade, and at what better time than to replace earthquake losses?

10

TOWARDS A SAFER ENVIRONMENT

INTRODUCTION

A decade for disaster reduction

Ninety-three countries have given their support to the International Decade for Natural Disaster Reduction (IDNDR), which began on 1 January 1990. The aim of this UN-sponsored decade is the creation and maintenance of a safe environment, and we would share in trying to achieve this goal. Yet the analysis of this book suggests a different approach from that of some of the work being done under the IDNDR. The main difference is in our analytical approach, which stresses that disasters are events that happen to vulnerable people. This shifts the emphasis away from the natural hazard itself towards the need for a much better understanding of the processes that generate vulnerability.

Much of the approach of the IDNDR is focused on the hazards themselves, and this runs the danger of encouraging top-down planning and mitigation, expensive prediction systems, and attempts to reduce hazards through technical measures that involve heavy capital expenditure. To none of these are we opposed altogether, but as the book should have demonstrated, they need not be appropriate, may reinforce the processes that generate vulnerability, and can replace one set of problems with another (including payment and indebtedness, which in many Third World situations is a burden that may fall disproportionately on the poor).

A major thrust of the IDNDR is to improve prediction, including weather prediction systems, earthquake and volcano monitoring, and understanding of the impact of hazard events on natural resources (NRC 1991: 4). We have discussed prediction and warning several times. Better scientific and financial resources to improve prediction are to be welcomed, especially when they are tied to improvements in warning systems. However, such work is at the 'prestige' end of the problem, in the domain of 'big science' (alongside the human genome project, the supercollider, and space programme). Will it overshadow the equally important work of understanding vulnerability from the bottom up?

Yet there are signs that the need for vulnerability analysis is being increasingly recognized at various conferences of the IDNDR at national and international level.[1] It is vital that this emphasis is recognized and reinforced, and that those involved in disasters work accept that the reduction of disasters is about reducing vulnerability, and that this involves changing the processes that put people at risk in addition to modifying the impact of the hazards themselves.

'Release' from the pressure

Our book argues that the environment cannot be made safer by technical means alone. The vulnerability perspective suggests both that it is possible to make the human environment safer and that there are limits set by the economic and social inequities, cultural biases, and political injustices in all societies. In this final chapter we pull together the experience reviewed in Part II and provide an overview of approaches to disaster mitigation and preparedness that aim at reducing vulnerability and creating safer environments.

The chapter has a number of objectives:

- to place vulnerability analysis and implementation within the broader context of disaster planning;
- to develop a series of principles concerning disaster risk reduction;
- to discuss some problems of terminology; and, finally,
- to link disaster mitigation with its wider context – the creation of a sustainable environment.

We are optimistic about the possibility of improvement. The 'pressure' model can be reversed to provide security instead of risk. Vulnerable people's access to resources can be improved and changes in power relations can be made. Vulnerability can be decreased, and if aid is properly conceived and implemented, even the most vulnerable survivors can recover in such a way that future vulnerability is reduced. Assessing risks must involve both the social vulnerability information and that concerning the natural hazard, using both sides of the 'pressure' model.

In this chapter we illustrate this by using a transformed version of the 'pressure' model used throughout this book, recalling that the 'R' in PAR model means 'release'. The outcome is 'safe' as opposed to 'unsafe' conditions, 'sustainable' versus 'unstable' or 'fragile' livelihoods, and 'resilience' or 'capable' versus 'vulnerable' people. Figure 10.1 summarizes the 'release' process as a reversal of disaster 'pressure'.

The social, economic, and political mechanisms (dynamic pressures) that translate global pressures into unsafe conditions for specific people can be blocked, changed, even reversed. To illustrate the way the 'pressure' process leading to disaster can be reversed, we should recall some of the 'good news'

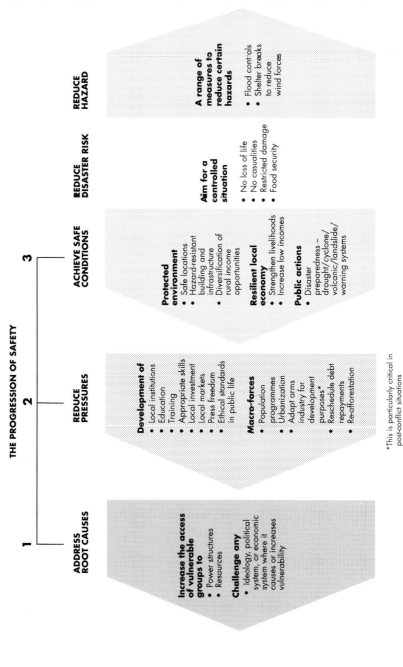

THE PROGRESSION OF SAFETY

| 1 | 2 | 3 |

1 ADDRESS ROOT CAUSES

Increase the access of vulnerable groups to
- Power structures
- Resources

Challenge any
- Ideology, political system, or economic system where it causes or increases vulnerability

2 REDUCE PRESSURES

Development of
- Local institutions
- Education
- Training
- Appropriate skills
- Local investment
- Local markets
- Press freedom
- Ethical standards in public life

Macro-forces
- Population programmes
- Urbanization
- Adapt arms industry for development purposes*
- Reschedule debt repayments
- Re-afforestation

3 ACHIEVE SAFE CONDITIONS

Protected environment
- Safe locations
- Hazard-resistant building and infrastructure
- Diversification of rural income opportunities

Resilient local economy
- Strengthen livelihoods
- Increase low incomes

Public actions
- Disaster preparedness – drought/cyclone/ volcanic/landslide/ warning systems

REDUCE DISASTER RISK

Aim for a controlled situation
- No loss of life
- No casualties
- Restricted damage
- Food security

REDUCE HAZARD

A range of measures to reduce certain hazards
- Flood controls
- Shelter breaks to reduce wind forces

*This is particularly critical in post-conflict situations

Figure 10.1 The release of 'pressures' to reduce disasters: the progression of safety

from chapters in Part II. Village-based and regional grain storage has successfully provided food reserves that rid some African farmers of the burden of indebtedness and the cycle of the 'hungry season' that can easily escalate into famine. Primary health care at neighbourhood and village level (including child immunization) has dramatically increased protection against some diseases, reducing vulnerability to other hazards, as has improved community water supply and sanitation.

Flood mitigation has also been achieved through local efforts, and the 'living-with-floods' approach is actively competing with technocratic views. Death and damage from tropical storms continues to be high. However, the grass-roots, decentralized efforts of Red Crescent workers in Bangladesh and the success of storm shelters encourage some optimism. Great progress has been made in low-cost housing design and construction that is much safer in earthquakes. Grass-roots efforts in this area are very impressive.

All these relate to the middle of the 'release' model. Such actions reverse the 'mechanisms' that translate global pressures into unsafe conditions. There has been less progress in relieving humans of the most pressing global pressures. Foreign debt, war, global environmental change, population growth, and urbanization still provide formidable challenges. However, there are moves that suggest that the influence of even these 'root causes' will eventually be decreased, such as debt forgiveness and rescheduling, and various debt-for-development swaps.

Root causes (global pressures) can be changed, and should not be regarded as immutable and inevitable. In its 1992 report on the state of the world's children, UNICEF recommended that most of Africa's external debts be forgiven (UNICEF 1992). Peace-keeping efforts by the UN and recent moves to assert the priority of humanitarian relief over national sovereignty suggest that progress can be made against the influence of war on disaster vulnerability. The UN Conference on Environment and Development focused attention on hundreds of ways sustainable livelihoods can be secured. Population growth and urbanization continue unabated. However, successful rural development seems to encourage people to have fewer children and also to stay in the countryside. Reductions in the other three global pressures (debt, war, environmental degradation) would allow more successful rural development, with potential benefits for a decline in birth rates and urban growth. The result would eventually be translated into less vulnerability to hazards.

On the side of the physical hazards, even these can be modified in many cases, though our concern is to ensure that mitigation and modification of the hazards are done in such a way that the investments and science involved do not create other forms of vulnerability, are really effective (in that, for instance, warnings are receivable and useful), and do not replace proper vulnerability analysis.

'Living with' hazards

Throughout our book we have mentioned local coping and the potential for building on it as a basis for vulnerability reduction. But we do not want to romanticize 'traditional' or 'local' coping, and consider it to be sufficient, 'the best', or the only kind of response to risk. We argue for a balanced approach that combines systems such as satellite warning networks and world stockpiles of food with national and local preparedness. In this context the mobilization of knowledge and efforts at the neighbourhood and village level are also critical. It would appear to be impossible to rid human existence of all risk: a completely safe environment is unattainable. Yet to exclude any of these various forms of vulnerability reduction, and to fail to recognize their interconnections, is potentially lethal.

The starting-point – not goal – is, therefore, the achievement of ordinary people in 'living with' floods, droughts, pest and plant disease attacks, epidemics, storms, earthquakes, steep unstable slopes, and volcanoes. Previous chapters have digested a wealth of evidence to show that people know a good deal about these hazards and are not passive victims.

MANAGING A REDUCTION OF VULNERABILITY

In the previous chapter on relief and recovery, we distilled the experience of planners and policy-makers into a series of guiding principles.[2] Governments are not the only agents that either increase or decrease disaster vulnerability (see Chapter 1), but policy at this level can have a large impact.

The term mitigation is commonly used in environmental studies and disaster management, but is not widely used outside this context. It can be defined as 'actions taken to prevent or reduce the risks from natural hazards' (NRC 1991: 3). Although it is often used to describe efforts to modify the actual hazard itself (through engineering or other forms of interference with the natural process), it is essential to see it in the broader context of reducing vulnerability in all its forms. The following principles are derived from the experiences of hazards we have discussed in this book, in conjunction with the processes that generate people's vulnerability through root causes and dynamic pressures. In other words, mitigation is not only about altering the hazard side of the PAR diagram, but must also be seen in the context of the progression of vulnerability.

Principle 1: Vigorously manage mitigation

Strong relief management in the moment of emergency is well understood, but it remains unusual for any government, ministry, agency, or even public official to have the overall responsibility for co-ordinating risk-reduction actions. Laws can be drafted by one sector of a government which bear no relationship to how they will be implemented by another, or for example, how they will be enforced, financed, or taught. Such laws, or codes of practice may even prescribe safety standards or building techniques which have still to be developed by yet another government agency. Therefore effective implementation of disaster mitigation requires strong management to integrate all elements into a cohesive pattern.

In Part II we saw that it has been recognized since the days of the British Raj in India that avoidance of famine involves the successful integration of many functions of government (Chapter 4). Faltering AIDS prevention was seen to be related to fragmented, non-integrated programmes (Chapter 5). Positive examples also included integrated governmental response to flood hazards in China (Chapter 6) and cyclone damage in India and Mozambique (Chapter 7). In Chapter 8 we learned how early commitment by the Mexican government to a comprehensive study of the 1985 earthquake and plans to decrease vulnerability were undermined by a change of regime.

Disaster mitigation also requires anticipation. Without this leadership and application of skills at many different levels of government and in the private sector, implementation will be slow and patchy. This is particularly true where the onset of the hazard is slow, and/or it recurs very infrequently (for example, drought, earthquakes, and eruptions by some volcanoes). There has to be a way of providing an institutional memory of hazard events and disasters that links with new generations of government administrators and planners and their bureaucratic culture and practice, and with the populace and its own collective memory preserved perhaps in popular culture (stories, songs, etc.) and practice (building, farming, etc.). Ideally a bridge would be built between popular commemorative and precautionary culture and its bureaucratic counterpart. This is not inconceivable, as our discussion of the samba schools of Rio de Janeiro will suggest later in this chapter.

One function of management will be to interrelate and even integrate structural and non-structural measures. For example, construction techniques (structural measures) and land-use planning controls (non-structural measures) may be developed for seismic zones. These will then need to be taught to urban planners, engineers, and builders. The actual construction techniques are obviously structural measures. Therefore there is a need for linkages in this 'chain' of safety measures involving legal frameworks, education, and implementation in an orderly, logical sequence of actions.

Management may well be self-defeating if it is based on a hierarchical model which is unresponsive to the needs of the staff within the relevant organization or the community they seek to serve. Implementation needs to be a balanced participatory system, related to the diversity of levels in the various sectors, ministries, and administrative structures, as well as in local communities who receive the safety measures.

Government officials and NGO workers must respect the knowledge and practice of ordinary people. As we saw in Chapter 3 and throughout Part II, people's coping mechanisms can be very effective. Spontaneous relief and recovery efforts can be very impressive, as the case of the Armenia earthquake showed; 90 per cent of the survivors were dug out of the rubble of buildings by other survivors. On the other hand, popular knowledge and practice can be flawed and inadequate (recall the building techniques described in the seismically dangerous zone of Pakistan in Chapter 8). Thus 'people's knowledge', 'folk ecology', or 'coping' should not be romanticized. It is a starting-point, not a definitive body of knowledge.

Principle 2: Integrate the elements of mitigation

The order in which risk-reduction measures are developed is of critical importance. Perhaps the ideal sequence would be: public awareness leading to political will, leading to management, leading to parallel and interactive processes of drafting laws. Finally, training and education and cash incentives will be required to apply such measures.

Many countries may already have some of these elements in place, and they may be highly effective. They can develop other protection elements to support and build upon existing strengths, and begin constructing mitigation measures which do not yet exist. Where there are such gaps, it is usually either because elite interests in the country or overseas benefit from the *status quo*, or because those who are specifically vulnerable are in various ways marginal (politically, economically, geographically isolated) and thus 'invisible' to national decision-makers.

Principle 3: Capitalize on a disaster to initiate or to develop mitigation

The major opportunity to develop or implement measures will occur in the wake of a disaster. This is due to the temporary high profile of disaster preventive action, which should be taken advantage of to secure resources and decisions. In Part II we saw that a variety of food security and relief systems (e.g. food-for-work) as well as innovations in public health, flood control, flood insurance, storm warning, and building safety have grown out of disaster occurrences.

Plans should be developed, and where there are political or other obstacles to their implementation, they should be maintained in readiness for implementation at the appropriate time, such as when a disaster provides the necessary window of opportunity for swift action. Such plans can be kept in readiness by middle-level administrators and planners who would normally fail to win the ear of directors. There could also be plans (e.g. for urban infrastructural lifelines) developed by mayors who would normally not receive national budget allocations allowing them to pursue such initiatives. Contingency plans can also be developed by non-governmental and popular organizations either before or after a disaster occurrence. In Chapter 8 we saw that survivors of the Mexico City earthquake refused to be relocated and countered with their own recovery plans.

Principle 4:   Monitor and modify to suit new conditions

This book has emphasized that, while the underlying causes of vulnerability and even global dynamic pressures are common, the particular risk can vary from situation to situation and can change rapidly with time. As particular patterns of vulnerability change, due to such pressures as urbanization and land degradation, then assessment techniques, implementation strategies, and mitigation actions also need to adapt flexibly.

Our book has offered vulnerability analysis as a complement to the physical mapping of hazards. Hazard-mapping is the assessment for the biological or physical trigger itself. Broadly speaking, 'vulnerability analysis' refers to everything else: monitoring changes in root causes (global pressures), and understanding how these are channelled into unsafe conditions for specific subgroups in the population by social and economic mechanisms (dynamic pressures).

Routine assessment activities already exist that focus on construction safety, harvest forecasts for staple crops (using both field observations and remote sensing), monitoring of the nutritional status of children, epidemiological surveillance, as well as numerous flood and storm warning systems. However, the surveys in Part II suggest that vulnerable people often suffer a series of interrelated disasters and that their vulnerability often increases through failure of recovery. Existing monitoring systems do not address these problems.

Our recommendation is that these existing routines incorporate an explicit vulnerability orientation and that their results (interpreted differently due to the vulnerability framework) become the focus of a co-ordinating governmental body ('disaster commission').

This suggests some quite straightforward guidelines for generating profiles of vulnerability and for monitoring changes in vulnerability. What remains is to integrate these simple procedures into the work of NGOs, government departments, and multilateral agencies on a trial basis as part of the International Decade for Natural Disaster Reduction (IDNDR).

The information needed is summarized in Table 10.1. It concentrates on five groups most likely to have least protection against hazards and least reserves for recovery. We have chosen these groups for purposes of demonstration only. Identification of vulnerable groups will vary from society to society, where very specific differences, based on caste, gender, class, age, may have a role. Our choices in this example are based on the common ground that the cases in Part II seem to suggest. Thus the five vulnerable groups would be:

1 the poorest third of all households;
2 women;
3 children and youth;
4 the elderly; and
5 some number of minority populations.

For each group we need to know several aspects of their 'access to resources' (as discussed in Chapter 3) as well as the hazardousness of the locations they frequent. In all cases we want to determine, at least in a qualitative way, the direction or trend over time. Are they gaining or losing access to resources that can help to protect them or help them to cope and recover? Are the spaces they inhabit becoming more or less hazardous? Have they been forced to move residence or job or have existing sites become more dangerous?

In Table 10.1 'natural resources' refers to land, water, forest products, and other productive assets. In many of the preceding chapters we have seen that the most vulnerable have been losing access to such resources over the past few decades. This is for many reasons, including privatization and concentration of land, increased inequality in income distribution, land degradation, and population growth.

'Physiological and social resources' refer to nutritional status and health, education, and access to technology and information. These change in positive or negative directions with time and across generations. For instance, in South Africa the social resource of kinship is reported to be less important to coping than it has been in the past (Sharp and Spiegel 1984). In the US some reports suggest literacy levels are actually declining with time in certain social groups. IMF and World Bank structural adjustment programmes have caused many governments to cut back on education and health budgets and staple food subsidies. This has acted as a 'root cause' of increased vulnerability because of the erosive effect it has on physiological reserves or resources.

'Financial resources' refers to income, market access, banking, and other credit facilities. Recalling the basic 'access' model from Chapter 3, our discussion of coping, and the debates surrounding famine in Chapter 4, it should be clear that the concept of 'financial resources' can be extended to include liquid assets that a household can sell to buy emergency food, health care, or to rebuild a house. They would include, for instance, livestock and jewellery. Overgrazing, privatization, or nationalization (as game reserves) can deprive herders of pasture, and so interfere with the assets people are accustomed to.

The last column contains information about the location of home and livelihood activities in relation to various hazard triggers. These include biological and physical variables affecting home and workplace such as slope, soil type, microclimate, and can be 'mapped'. These interact with

226

known (or knowable) risks such as that of torrential rain or the introduction of cholera into the locality. Also relevant is the presence or absence of toxic risks, or the potential for catastrophic industrial accidents (although here we come up against the limit of our self-imposed scope). These 'hazard maps' for vulnerable groups change over time as human groups are forced to move about the landscape and as the spatial distribution of livelihood opportunities changes.

In this systematic manner it is possible to gather, display, and interpret information of very diverse kinds that may influence the vulnerability to disaster of a specific subgroup in society.

Table 10.1 Types of information required for vulnerability analysis

Potentially vulnerable group	*Natural resources*	*Physiological and social resources*	*Financial resources*	*Hazardousness of home and workplace*
Poorest 33%	+ 0 −	+ 0 −	+ 0 −	+ 0 −
Middle 33%	+ 0 −	+ 0 −	+ 0 −	+ 0 −
Richest 33%	+ 0 −	+ 0 −	+ 0 −	+ 0 −
Women	+ 0 −	+ 0 −	+ 0 −	+ 0 −
Children	+ 0 −	+ 0 −	+ 0 −	+ 0 −
Elderly	+ 0 −	+ 0 −	+ 0 −	+ 0 −
Minority Group A	+ 0 −	+ 0 −	+ 0 −	+ 0 −
Minority Group B	+ 0 −	+ 0 −	+ 0 −	+ 0 −
Minority Group n	+ 0 −	+ 0 −	+ 0 −	+ 0 −

Notes: In each cell of the table, the symbols indicate the trend in the data for each group and type of data: + indicates increase, 0 indicates no change, and − means a decline. For each group of people, their trend of vulnerability (up, unchanged, or down) will be the amalgamation of the data in that particular row.

Trends could be established by inquiring about each of these items, by row and column, in regard to the situation two generations ago, ten years ago, and at present. The basic question in each case is whether the specific resource or locational factor is significant for the increasing vulnerability of a specific group to a hazard, and whether this factor is enhancing vulnerability more or less over time.

Most countries have sufficient data to identify such groups sufficiently to allow statistically significant sampling. The World Bank and the UN Institute for Research in Social Development (UNRISD) have produced a massive literature on how to define and to measure poverty (e.g. McGranahan, Pizarro, and Richard 1985; Lipton 1988). Having established society-specific definitions and a sampling framework, the data could be collected in a number of ways including conventional surveys and participatory action research (Kalyalya *et al.* 1988; Wisner, Stea, and Kruks 1991) where NGOs have broad-based programmes.

'Action research' is investigation in which people are co-researchers and the goal is practical improvement and policy change. It is underutilized in disaster planning, relief, and mitigation (Anderson and Woodrow 1989; Maskrey 1989), but has shown great potential in fields related to vulnerability reduction such as rural energy (Gamser 1988), farming (Chambers, Pacey, and Thrupp 1989), indigenous forestry (J. Clay 1988), water supply and sanitation (A. White 1981), and housing (Turner 1982).

Some fieldwork has already been carried out along these lines. In Chapter 4 we mentioned some experiments with decentralized early warning systems for famine in Africa, carried out by international NGOs such as Save the Children and the Red Crescent Society, some of them involving ordinary citizens (Cutler 1984, 1985; York 1985).

A similar, though retrospective, study of socio-economic vulnerability was carried out by UNICEF in Mozambique (D'Souza 1988). In this case quite specific characteristics marking vulnerability were revealed: old age, lack of access to low-lying land or traditional water-lifting devices, few family members working abroad in South Africa or insufficient money sent home from those who were, isolation from health centres, long distance to farms, women too busy to take children to the health centre (D'Souza 1988: 32). It is striking how well these characteristics fit into Table 10.1. Access to resources and technology are represented, as are income, age, gender, and location in relation to services and elements of the total livelihood system (farms, long-distance employment in South Africa).

Principle 5: Focus attention on protection of the most vulnerable

The priority is for appropriate measures to protect the most vulnerable groups of people. This book has shown that it is often (though not always) the poor who are most vulnerable to hazards, along with others that are vulnerable because of some combination of class, gender, age, and ethnicity.

Such measures must include economic improvement in access to resources of various kinds for the most vulnerable groups. These access issues can be read directly from Table 10.1 and were discussed in more depth in Chapter 3. Access to land, to water, to trees and other forms of biomass, to wild genes is crucial for many highly vulnerable rural people. Common property resource (CPR) systems do a great deal to improve access of the resource-poor (Chambers, Saxena, and Shah 1990; Jodha 1991). However, in many parts of the world access to CPR is declining as such assets are privatized or enclosed by more powerful groups. Adequate access to animal traction and labour is often a constraint that increases the vulnerability of the rural poor, especially women on their own. Nutritional supplements (for children and lactating mothers, for instance) can do a great deal to increase physiological reserves to withstand crises. Social resources include local organizations (co-operatives, self-help groups, churches, mosques, and temples) as well as educational and health infrastructures. These appeared in a positive light again and again in Part II. Such basic social infrastructure provides a means for communicating disaster mitigation ideas, as well as physiological support and skills (in sanitation, literacy, etc.) that allow more effective self-protection and more rapid recovery by survivors.

Access to financial resources can also be improved. However, this need not be thought of as a government-sponsored 'give away' programme. New banking systems for women are springing up in Asia (e.g. the Gramin Bank in Bangladesh), Africa, and Latin America. Rotating loan funds, livestock restocking schemes, village and regional 'grain' and 'seed' banks are all possible arrangements based on success in various parts of the world. Grain 'banks' free farmers from having to sell their harvest cheaply and buy it back at many times the price during a crisis or even a 'normal' hungry season (see Chapter 4).

Principle 6: Focus on the protection of lives and livelihoods of the vulnerable

Where resources for mitigation are severely limited, it is vital to focus on the protection of lives, rather than giving priority to the protection of property. However, this has to be applied in a broader context, since the protection of some aspects of property, namely those related to livelihoods and income generation, is of critical importance in protecting lives in the long run.

Contrary to the perceptions of elite groups, the property of the poor is often more valuable in this respect than that of the rich. Also as seen in Chapter 7, the rich in Andhra Pradesh lost more property due to cyclone damage than the poor, but they were able to re-establish their economic stability far sooner. Poor people's homes may also be their workshop and warehouse. A few agricultural implements or an old boat are, to the poor, the means of subsistence. Where resources allow protection of property belonging to the affluent in society, priority should be given to factories, farms, mines, and plantations that provide employment.

Principle 7: Focus on active rather than passive approaches

'Passive measures' include the use of laws, codes of practice, or planning controls. While they may be reasonably effective in wealthy, industrialized countries, they can be less reliable in the Third World. In contrast, 'active measures' would include the use of incentives and the sort of resource transfers mentioned above (Principle 5), training programmes, and institution building (especially at the grass roots). Experience suggests that they are much more likely to succeed with poor people than passive approaches.

The legal apparatus of mitigation (building codes, flood plain land-use zoning, crop insurance, etc.) is evolving in all parts of the world, and will continue to do so. But priority for financial and other support of grass-roots organizations has a greater significance. Village and neighbourhood organiz-ations are more likely to lobby so that legislators enact the passive (legal) initiatives in favour of the most vulnerable. Otherwise laws and regulation that tend to emerge in class society benefit the rich and powerful (e.g. large landowners, middle-class home-owners).

Principle 8: Focus on protecting priority sectors

Mitigation should focus on overall protection of an entire community and its property. However, with limited resources and unequal patterns of vulner-ability, the risk assessment and planning processes should identify priorities for mitigation measures. These should first address the needs of the vulner-able and poor. Secondly, they should be based on criteria generated in a debate that is specific to the political and cultural situation in each country. Likely priorities include:

- maximum number of people to be protected for given resources (i.e., protecting multi-occupancy buildings rather than individually-occupied dwellings);
- 'lifeline' services (i.e., water, sanitation, medical facilities, fire protection, communication systems, etc.);
- elements of long-term economic importance rather than short-term (i.e., factories before shops, certain trees such as coconuts that take years to replace before annual crops);
- food stocks;
- cultural monuments and artifacts since their protection may be a critical factor in the morale of the community during recovery (i.e., historical buildings, sculpture, paintings, books, museum objects (see Oliver-Smith 1986b)).

Since priorities must emerge from a national debate, it is vital to empower the vulnerable through their grass-roots organizations and non-governmental organizations so their voices are heard (compare Principles 5, 7 and 12).

Principle 9: Measures must be sustainable over time

It is one thing to initiate mitigation, but quite another to sustain it over time until it is fully incorporated into budgets and planning procedures. Ways to maintain mitigation include:

1 an active public awareness programme;
2 well-documented success stories of risks that *were* reduced in a past disaster;
3 institutionalizing mitigation into normal governmental planning and policy;
4 good use of expanded cost–benefit analysis to show the gains from protection.

On average in the Third World the cost of disaster (the sum of direct relief, reconstruction, lost production, etc.) can amount to as much as 5 per cent of GNP (Wisner 1976b; Burton, Kates, and White 1978). When corporations, large landowners, and government leaders are aware of such figures, they often see the sense of investments in mitigation. However, these programmes are often of a one-off variety, especially when initiated in the wake of a disaster (see Principle 3). The recurrent expenses of maintaining lifeline infrastructure and supporting improved assess to resources for the most vulnerable must be maintained. What these sources should be in a given nation (taxes, levies on certain kinds of non-renewable resource exploitation, etc.) has to be the subject of a national debate in which the voices of the vulnerable must be heard.

Principle 10: Assimilate mitigation into normal practices

A successful strategy will incorporate risk-reduction measures into normal practice. Put another way, risk-reduction needs to be absorbed into the developmental programme of any hazard-prone Third World country. Some examples were given in Part II. For instance, Botswana's famine early warning system is part of the 'normal' function of child health operations; the Indian meteorological service has a 'routine' storm warning procedure.

In industrialized countries architects, engineers, builders, home-owners and occupants have become so familiar with fire-resistant building materials that they virtually cease to notice their existence. This may have the negative impact of creating a false sense of security, but in positive terms it means that fire protection has been absorbed into building practice and public awareness. A parallel process is the way preventive medicine has been gradually accepted as a normal process of public health care.

Therefore preventive risk-reduction measures are best implemented by getting them incorporated into the culture which embraces government structures, traditions, curricula, laws, training schemes, political practices, and public awareness. The long-term planning undertaken in many countries (such as Five Year Economic and Social Development Plans) is a good vehicle for systematic disaster mitigation. The plans need a disaster vulnerability chapter, and hazard mitigation needs to be one of the drafting guidelines provided by co-ordinating authorities to all sectoral working parties. Vulnerability analysis and reduction can also be incorporated into normal routine through population censuses. These offer a perfect opportunity for gathering and updating precisely the information called for in Table 10.1 above (Principle 4).

Principle 11: Incorporate mitigation into specific development projects

Just as major development projects are checked for their environmental impact by means of 'environmental impact assessment' it is vital that a similar check be made on the project's consequences for disaster vulnerability. The aim of such a 'hazard impact analysis' would be:

- to verify that new projects will not increase the risks of such hazards as flooding, landslides, soil erosion, desertification, or disease;
- to protect the investment from being damaged or destroyed in a future disaster.

A number of authors have suggested concrete ways of including ecological impacts in project assessment (Bartelmus 1986: 62–81; Dixon *et al.* 1988; Pearce, Markandya, and Barbier 1989; Pearce, Barbier, and Markandya 1990; Pearce and Turner 1990). It should be relatively easy to extend project assessment techniques to include disaster vulnerability. The conceptual framework provided by the notion of 'sustainable development' (WCED 1987b) gives such efforts by planners, officials, and citizen lobbyists technical legitimacy, and major international events such as the Rio Summit (UNCED) and the IDNHR itself provide political legitimacy.

Principle 12: Maintain political commitment

Without strong popular political pressure and freedom of expression in a given country, risk-reduction measures are likely to be at best token responses. Political will is most likely to originate from a major failure to deal with a disaster (Principle 3). But such a 'catalytic' occurrence and the existence of responsible and concerned officials is not enough. Just as occupational and public health advances of the nineteenth century were the result of popular *demands*, reduction of disaster vulnerability must become a demand of workers, peasants, students, and urban 'squatter citizens', organized in such a way that their interests cannot be ignored in government decision-making, budgeting, and project assessment. A further condition necessary for the incorporation of these opinions and voices into a national debate is democracy. We return to this subject at the end of this chapter.

In the final sections of this chapter, we turn to the *content* of disaster prevention and mitigation measures. From what repertory of policy measures can planners choose? What kinds of specific measures could become popular demands during the decade of disaster reduction?

DISASTER PREVENTION AND MITIGATION

In the early 1970s when disaster planning was being formulated into operational management tools, the rather optimistic term 'disaster prevention' was used very freely. With hindsight the term has a decidedly utopian ring, and by 1991 UNDRO had stated about prevention: 'The term should not be used as it implies misleading resource allocation. It is false to suggest that infinite risk can be matched by infinite resources' (UNDRO 1991: 157). Given the acute resource limitations that prevail throughout the Third World, the more realistic term 'mitigation' has replaced 'prevention' in recent years. An editorial of the *UNDRO News* in 1989 suggested that

> the reason for the evermore widespread use of the term mitigation seems to be its inherent sense of realism. Prevention has sometimes been found misleading, inasmuch as many disasters cannot be prevented from taking place. Mitigating, that is to say damping, the worst effects of violent and sudden natural hazards, is well within the realm of reality and also within the means of most developing countries.
>
> (UNDRO 1989: 2)

This is an excellent description of mitigation, which can be defined in an abbreviated form as 'actions taken to reduce damage and loss'. There is a series of distinct ways in which this can take place, and risk-reduction typologies abound (G. White 1974; Burton, Kates, and White 1978; Petak and Atkisson 1982; Drabek 1986; Palm 1990). These include measures to reduce the physical hazard, to provide structural and non-structural mitigation (including specialized agricultural mitigation measures), and to increase preparedness. Our particular contribution is to suggest that mitigation can be viewed from the point of view of vulnerability, vulnerability reduction, and popular attempts at coping and self-help.

CONCLUSION:
TOWARDS SUSTAINABLE REDUCTION OF DISASTERS

The discussion in this book has revealed a marked tendency in conventional disaster work to treat symptoms rather than causes. The reason for this bias is because vulnerability is deeply rooted, and any fundamental solutions involve political change, radical reform of the international economic system, and the development of public policy to protect rather than exploit people and nature.

The authors believe that creating a safer environment is basically an ethical concern directed to those with power, who have opportunities to make the world safer for the vast majority who are vulnerable because they are subordinated and unable to make choices that make them safe. Maskrey has observed that the only 'choices' available to the residents of Lima's squatter settlements were 'between different kinds of disaster . . . people seek to minimise vulnerability to one hazard even at the cost of increasing their vulnerability to another' (Maskrey 1989: 12).

The United Nations Development Programme has recently defined a 'Human Development Index' as an alternative to the customary GNP per capita index of national development (UNDP 1990a). This emphasized the importance of 'choice', and in fact defined human development as 'a process of enlarging people's choices'. But the harsh reality is that increasing choice within a society involves political change, economic reform, social compassion, and 'changes of heart'. While these are the essence of political rhetoric, they are rarely the priorities for policies or funding at either national or international levels. Governments invariably operate with short-term perspectives, many with the desire to win the next election.

It becomes clear that there are some formidable obstacles to the success of the 1990s as a decade for disaster reduction. The editor of the *New Scientist* put the issue very bluntly:

> All the aims of the IDNDR will cost money, and, in particular, money for things that appear to have no immediate benefit. Few politicians will support something that could bring no visible benefit for 10 or 20 years, let alone a century. Add to this the fact that many of the measures that could cut the death toll from disasters will disrupt people's lives, and you have a very good excuse for doing nothing.
>
> (*New Scientist* 1989: 3)

In preparing for the IDNDR, an international committee of experts anticipated this difficulty for governments in 'selling' protection policies to their constituencies. It therefore proposed that: 'efforts to reduce the impacts of natural disasters will gain far wider acceptance if they are perceived as a means to protect economic development and improve living standards rather than to mitigate some hypothetical, localised and infrequent event' (Austin 1989: 65). There are parallels here with the development of 'green politics', where certain governments are being persuaded by carefully-framed economic arguments to make policy reversals often against powerful vested interests. These are also long-term measures that cost large sums of money and cause disturbance.

However, while there is now a global lobby that has become a political force to apply the pressure for environmental protection, no such lobby yet exists for protection from disasters. Also environmental demands have widespread popular support in many countries where millions of ordinary people have returned 'green' candidates to city councils, state and national parliaments. Disaster vulnerability has not yet become so widely accepted as a popular issue, partly because of the persistence in the media and education system of the myth of natural causation of disasters.

At the political level, then, reduction of disaster vulnerability has got to become a 'green' issue. This may require more work in drawing out the similarities of vulnerability to 'natural' disasters to the range of 'technological' disasters about which 'green' movements are highly conscious, such as Bhopal, Chernobyl, Sellafield, Three Mile Island, Love Canal, and Exxon Valdez. To date there has been little contact between academics and activists working in these two areas. Links must also be made between the various bureaucracies that deal separately with disaster risk-reduction and environmental protection.

Achieving radical changes to address the fundamental causes of vulnerability is extremely difficult. The mitigation strategy outlined in this chapter therefore focuses on policies and procedures that will reduce some risks even if they leave many vulnerabilities resolutely intact. It is perhaps best described as the 'art of the possible'. The more difficult measures that will be needed to reduce vulnerability significantly involve changes in power relations and economic systems. The struggle to achieve better protection for the vulnerable will in itself alter these relations, and demonstrate to dominant groups and elites that shared solutions to vulnerability are worthwhile.

Popular grass-roots involvement in assessing risks and in designing and implementing mitigation measures can have the further, long-term effect of giving people the self-confidence and organization to demand more. In the words of Boyden and Davis:

> disaster mitigation has implications which are quite different – and much further-reaching – than those of disaster relief. First, relief by its very nature creates dependency between donor and recipient. Mitigation on the other hand aims to increase the self-reliance of people in hazard-prone environments to demonstrate that they have the resources and organisation to withstand the worst effects of the hazards to which they are vulnerable. In other words, disaster mitigation – in contrast to dependency creating relief – is empowering.
>
> (1984: 2)

Popular and professional experience with disasters has now accumulated to the point that in the 1990s reducing vulnerability is within the bounds of imagination. The obstacles to disaster impacts being considerably reduced are less to do with the available knowledge than the development of people's demands, of supporter's actions, and of corresponding political and economic shifts.

Action in the fields and on the streets: tapping the energy of citizen–activists for disaster mitigation

Critics of *status quo* development efforts have for some time called attention to the potential of what they call the 'third system'. This means ordinary members of civil society who organize to demand their rights and to protect their communities (Nerfin 1990). The best hope for the Disaster Decade could well be an upsurge of such citizen action. In other contexts this phenomenon has been very powerful. The worldwide movement of consumers' unions has successfully confronted the power of the pesticide and baby food industries. In the US, activist Lois Gibbs (who saw her family sicken in her home at a place called Love Canal) now co-ordinates some 7,000 citizen groups who make up a national toxic monitoring network. Similarly, survivors of Chernobyl have become part of the rapidly growing citizens' environmental movement in the territories of the former Soviet Union. Survivors of the poisoning from the Bhopal chemical factory explosion have also become activists. There are 4,000 member-organizations of the Environmental Liaison Centre (based in Nairobi), and the majority are in the Third World.

In many parts of the world, school-leavers and university graduates are returning to their home communities to use their knowledge. They struggle alongside neighbours to reclaim sustainable agricultural technologies half-forgotten or buried by the influence of the Green Revolution. They fight against 'mega-projects' such as large-scale dams in India, Brazil, Mexico, the Philippines, Canada, and elsewhere. In other urban and rural communities, churches and other bodies form the centres for citizen response to economic dislocation and crisis. Food banks, community kitchens, and pantries have sprung up all over the US and in many Latin American countries to assist and involve poor and hungry people. People's health centres and public health movements have also emerged in the slums of many of the world's mega-cities from Brooklyn and the Bronx to Rio de Janeiro, Mexico City, and Manila.

Such formal and informal organizations are woefully underutilized by authorities responsible for disaster mitigation. Non-governmental organizations have been quicker to recognize the potential of such groups. The people themselves can campaign to secure their livelihoods and life spaces (Anderson and Woodrow 1989), to recognize the 'untapped power of people's science' (Wisner, O'Keefe, and Westgate 1977), and the effectiveness of 'community-based' mitigation (Maskrey 1989). People organize themselves spontaneously in many areas of life. Often such organizations as tenant unions, squatters' councils, and purchasing co-operatives arise as a way of protecting individually less powerful people from the power of landlords, urban officials, or profiteering retailers. These organizations already exist for reasons of self-protection and perceived threats in the local environment.

236

In such cases it is quite typical for them to turn their attention to the physical hazards that often affect them. In Rio de Janeiro, for instance, arguably the most powerful centres of community organization in the *favelas* neighbourhoods are the samba schools. Each year they organize dancers, make elaborate costumes, and plan for months in advance of the great carnival parade and competition. They are a focus of poor people's hope, pride, and energy. Could the samba schools become a major force for reducing vulnerability to urban floods and mudslides in Rio? They could be ideal organizations to press municipal authorities to clear the rubbish that often blocks drainage and leads to floods (see Chapter 8). In fact, in at least one of Rio's *favelas*, a grass-roots organization that began campaigning for public health improvements has taken on the issue of floods (SINAL 1992; Williams 1992).

Such popular organizations – as diverse as trade unions, consumers associations, churches, and samba schools – can contribute in the entire range of possible preventive and mitigating action. The most effective protection for trees and watersheds (and hence their livelihoods) in northern India has come from the village-based Chipko movement (Shiva 1989). This activity clearly falls under the heading of physical hazard modification. Is it not a case of structural mitigation when neighbourhood groups rebuild their houses using low-cost improvements to make them safer? (Maskrey 1989; Aysan and Davis 1992). Women's revolving loan funds in southern Zimbabwe and indigenous livestock loan arrangements in Mali must count as non-structural mitigation in the face of such hazards as drought, epizootic disease, and plant pests (Wisner 1988b).

However, two important things need to be said about this vast, heterogeneous 'third system'. Firstly, the 1980s and 1990s have produced a number of NGOs that do not represent popular constituencies. They are either creations of foreign NGOs or are run by a self-serving elite which has found a new gambit in the 'development game' (Hancock 1989). In some countries 'big men' have been known to use women's self-help groups as fronts for their economic ventures. Such groups are usually easy to spot. They should be judged by their actions, not their rhetoric. As noted in Chapter 9, accountability is the key.

Secondly, authentic NGOs are sometimes the target of government repression. Citizen architect–builders who were trained by Oxfam America in the aftermath of the Guatemala earthquake in 1976 were actually assassinated by death squads, as were rural health workers. The elite felt increasing pressure for land reform around 1980, and such community workers and NGOs were judged to be part of the threat. This sobering history must be kept in mind. International networks for publicizing abuse of the human rights of such activists exist, and will need to extend their support to those who would demand reduction of risk from hazards.

Ultimately it is only democracy and human rights that can ensure the positive role of citizen mobilization for risk-reduction. Officials should not have to criticize their nation's disaster policies from exile (as did the former Commissioner of Ethiopia's Relief and Rehabilitation Commission, Dawit Wolde-Giorgis). Famines should not have to be uncovered with the help of foreign authors because the governments are ignoring them (Clay and Holcomb 1985; Clay, Steingraber, and Niggli 1988; Article 19 1990). The 1990s has a better chance of seeing a successful decade of disaster reduction if it also lives up to the promise of a decade of democracy and human rights.

But we must look very carefully and critically at what is applauded as 'democracy' and whether it means more than simply voting every few years. In recent years there have been very significant changes in the former Soviet Union and Eastern Europe, moves towards multi-party, civilian governments throughout Africa, the overthrow of long-standing dictatorships in Haiti, the Philippines, and elsewhere, the return to civilian government in Nigeria, Bangladesh, and Pakistan. These do not necessarily reduce the vulnerability of ordinary people, though they may make it easier to campaign for it.

Reducing vulnerability also means enabling people to have access to safer residential locations (not in a ravine, steep hillside, or flood plain); it involves use of lighter, stronger, fire-resistant building materials; entitlements to a diet that will provide resistance to disease. Reduction of disaster vulnerability requires full, day-to-day participation of ordinary people and their own popular organizations in the struggle to enlarge choice and reduce vulnerability. This participation must be asserted and protected as a right.

There must also be a certain minimum level of social peace and need-satisfaction before many of the most vulnerable can take the chance to become publicly involved. In war-torn countries such as Sudan, Mozambique, El Salvador, or Lebanon it is virtually impossible to find the time required for co-operative effort to deal with natural hazards. Lives are stretched to the limit by the struggle to make ends meet, and deal with the continuing disaster of war. Where they have jurisdiction, governments must be responsible for constitutional protection of rights, and for a safety net for basic need-satisfaction sufficient to allow the most vulnerable to participate (Doyal and Gough 1991; Wisner 1988b). Is that asking so very much of governments? If the nation state has any moral ground for legitimacy, it must be the provision of such basic protection and empowerment. Where war is making national government jurisdiction ambiguous, difficult, or impossible, new forms of international intervention are needed not only to bring an end to the hostilities, but to ensure that the peace is constructed with a built-in concern for vulnerability to disasters. We hope this book will provide a source of analysis and information to assist people's demands for the right to a safe environment, and a tool to promote the reduction of vulnerability.

But war is not the only global pressure that we have argued lies at the root of the chain of disaster causation. There are other root causes that are less intractable, for which there is less excuse for inaction or inadequate responses. If our analysis in this book is persuasive, it must lead to the understanding that disasters are reduced only by releasing people from the unsafe conditions that derive from social, economic, and political pressures. The connections between vulnerability and these pressures need to be confronted, and the links to deeper root causes given the same priority for research and understanding as is available to the scientific and technical approaches. Then policies can be changed and resources can flow to deal with these causes, in addition to the current approach which emphasizes predicting or modifying hazards with generally top-down strategies and expensive projects.

NOTES

1 See, for instance, the programme for the International Conference in Yokohama in 1994, with a full session on Social Vulnerability Assessment. British IDNDR conferences in 1993 likewise included sessions on vulnerability and vulnerable communities.

2 These mitigation principles have been expanded from the version that first appeared in Davis and Gupta (1991).

BIBLIOGRAPHY

Abbot, S. 1991. Courting Ruin: Disaster Vulnerability. *Geographical Magazine* August: 12–15.

Abu Sin, M. and Davies, H.R.J. 1991a. Greater Khartoum's Vulnerability to Disaster Hazards: The Case of 1988 Rains and Nile Flood. In: M. Abu Sin and H. Davies (eds), *The Future of Sudan's Capital Region: A Study in Development and Change*, pp. 120–31. Khartoum: Khartoum University Press.

—— (eds) 1991b. *The Future of Sudan's Capital Region: A Study in Development and Change*. Khartoum: Khartoum University Press.

Achebe, A., Hyden, G., Magadza, C., and Pala Okeyo, A. (eds) 1990. *Beyond Hunger in Africa*. Nairobi and London: Heinemann Kenya and James Currey.

Adamson, P. 1982. The Rains. In: J. Scott (ed.), *The State of the World's Children 1982–83*, pp. 59–128. New York: UNICEF.

Adedeji, A. 1991. Will Africa Ever Get Out of its Economic Doldrums? In: A. Adedeji, O. Teriba, and P. Bugembe (eds), *The Challenge of African Economic Recovery and Development*, pp. 763–82. Oxford: Oxford University Press.

Adnan, S. 1993. Social and Environmental Aspects of the Flood Action Plan in Bangladesh: a Critical Review. Paper presented at Conference on the Flood Action Plan in Bangladesh, European Parliament, Strasbourg, May.

Agarwal, A., Kimondo, J., Moreno, G., and Tinker, J. 1989. *Water, Sanitation, Health – For All?*. London: International Institute for Environment and Development/Earthscan.

Agarwal, B. 1986. *Cold Hearths and Barren Slopes*. New Delhi and London: Allied Publishers and Zed Press.

—— 1990. Social Security and the Family: Coping with Seasonality and Calamity in Rural India. *Journal of Peasant Studies* 17,3: 341–412.

Akong'a, J. 1988. Drought and Famine Management in Kitui District, Kenya. In: D. Brokensha and P. Little (eds), *Anthropology of Development and Change in East Africa*, pp. 99–120. Boulder, Colo.: Westview.

Alamgir, M. 1980. *Famine in South Asia*. Cambridge, Mass.: Oelgeschlager, Gunn, & Hain.

—— 1981. An Approach Towards a Theory of Famine. In: J. Robson (ed.), *Famine: Its Causes, Effects and Management*, pp. 19–44. New York: Gordon & Breach.

Alexander, D. 1985. Death and Injury in Earthquakes. *Disasters* 9,1: 57–60.

—— 1989. Urban Landslides. *Progress in Physical Geography* 13,2: 157–91.

Ali, A.M.S. 1987. Intensive Paddy Agriculture in Shyampur, Bangladesh. In: B. Turner and S. Brush (eds), *Comparative Farming Systems*, pp. 276–306. New York: The Guilford Press.

Ali, M. 1987. Women in Famine. In: B. Currey and G. Hugo (eds), *Famine as a Geographical Phenomenon*, pp. 113–34. Dordrecht: D. Reidel.

Ali, T. 1982. The Cultivation of Hunger: Towards the Political Economy of Agricultural Development in the Sudan 1956–1964. Ph.D. Thesis, University of Toronto.

Allan, W. 1965. *The African Husbandman*. London: Oliver & Boyd.

Allen, E. 1994. Political Responses to Flood Disaster: The Case of Rio De Janeiro, 1988. In: A. Varley (ed.), *Disasters, Development and the Environment*. London: Belhaven.

Altieri, G. 1987. *Agroecology*. Boulder, Colo.: Westview.

Alvares, C. and Billorey, R. 1988. *Damming the Narmada*. Penang, Malaysia: Third World Network/Asia–Pacific People's Environment Network.

Ambraseys, N.N. 1988. Unpublished Notes of a Presentation on the Mexican Earthquake of 1985 to a Workshop on Disaster Management 4–5 July, p. 2. Oxford: Disaster Management Centre.

Amin, S. 1990a. *Delinking: Towards a Polycentric World*. London: Zed Press.

—— 1990b. *Maldevelopment: Anatomy of a Global Failure*. Tokyo and London: United Nations University Press and Zed Press.

Anderson, J.N. 1987. Lands at Risk, People at Risk: Perspectives on Tropical Forest Transformations in the Philippines. In: P. Little and M. Horowitz (eds), *Lands At Risk*, pp. 249–68. Boulder, Colo.: Westview.

Anderson, M. 1990. Which Costs More: Prevention or Recovery? In: A. Kreimer and M. Munasinghe (eds), *Managing Natural Disasters and the Environment*, pp. 17–27. Washington, DC: World Bank.

Anderson, M.B. and Woodrow, P.J. 1989. *Rising from the Ashes: Development Strategies in Times of Disaster*. Boulder, Colo.: Westview.

Anderson, R.M. and May, R.M. (eds) 1982. *Population Biology of Infectious Diseases*. Berlin: Springer-Verlag.

Andrae, G. and Beckman, B. 1985. *The Wheat Trap: Bread and Underdevelopment in Nigeria*. London: Zed Press.

Annis, S. 1988. What is Not the Same About the Urban Poor: The Case of Mexico City. In: J.P. Leavis (ed.), *Strengthening the Poor: What have we learned?* Washington, DC: Overseas Development Council.

Anton, P., Arnold, K., Truong, G., and Wong, W. 1981. Bacterial Enteric Pathogens in Vietnamese Refugees in Hong Kong. *Southeast Asian Journal of Tropical Medicine and Public Health* 12: 151–6.

Applebome, P. 1989. After Hurricane, Relief Gives Way to Despair. *New York Times*, 27 September, p. A1.

Arante, R.A. (n.d.). *Taal Volcano*. Quezon City: Philippine Institute of Volcanology and Seismology.

Arante, R.A. and Daag, A.S. (n.d.) (c. 1988) Evacuation Scheme for Taal Volcano. Unpublished Terminal Report submitted to the Geologic Disaster Preparedness and Planning Division (GDAPPD), of the National Disaster Co-ordinating Council (NDCC). Manila: Government of the Philippines.

Armenia Earthquake Reconnaissance Report. 1989. Special issue of *Earthquake Spectra* August. Edited by L.A. Wyllie and J.R. Filson. See especially Chapter 10 by F. Krimgold.

Armstrong, W. and McGee, T.G. 1985. *Theatres of Accumulation: Studies in Asian and Latin American Urbanization*. London: Methuen.

Arnold, D. 1988. *Famine: Social Crisis and Historical Change*. Oxford: Blackwell.

Article 19 (ed.) 1990. *Starving in Silence: A Report on Famine and Censorship*. London: Article 19 (International Centre on Censorship).

Asefa, A. 1986. The Ethiopian Famine. Unpublished manuscript.

241

Austin, T. 1989. Decade for Natural Disaster Reduction. *Civil Engineering* December: 64–5.

Aykroyd, W. 1974. *The Conquest of Famine*. London: Chatto & Windus.

Ayres, R.L. 1983. *Banking on the Poor: The World Bank and World Poverty*. Washington, DC: Overseas Development Council.

Aysan, Y. and Davis, I. (eds) 1992. *Disasters and the Small Dwelling: Perspectives for the UN IDNDR*. London: James & James Science Press.

Aysan, Y. and Oliver, P. 1987. *Housing and Culture after Earthquakes – A Guide for Future Policy Making in Seismic Areas*. Oxford: Oxford Polytechnic.

Aysan, Y.F., Coburn, A.W., Davis, I.R., and Spence, R.J.S. 1989. *Mitigation of Urban Seismic Risk: Actions to Reduce the Impact of Earthquakes on Highly Vulnerable Areas of Mexico City*. Report of Bilateral Technical Co-operation Agreement between the Governments of Mexico and the United Kingdom. Oxford and Cambridge: Disaster Management Centre, Oxford Polytechnic, and University of Cambridge.

Bach, W. 1990. Panel presentation at the session Sharing the Global Village, G.F. White, chair Annual Meeting of the Association of American Geographers/ Canadian Association of Geographers, Toronto, 21 April.

Baird, A., O'Keefe, P., Westgate, A., and Wisner, B. 1975. *Toward an Explanation of Disaster Proneness*. Occasional Paper No. 11. Disaster Research Unit, University of Bradford.

Ballard, P. 1984. The Miskito Indian Controversy. *Antipode* 16,2: 54–64.

Barker, D. and Miller, D. 1990. Hurricane Gilbert: Anthropomorphizing a Natural Disaster. *Area* 22,2: 107–16.

Barnaby, F. (ed.) 1988. *The Gaia Peace Atlas*. New York: Doubleday.

Barnett, A.S. and Blaikie, P.M. 1989. Aids and Food Production in East and Central Africa: A Research Outline. *Food Policy* 14,1: 2–7.

—— 1992. *Aids in Africa: Its Present and Future Impact*. London: Belhaven.

Bartelmus, P. 1986. *Environment and Development*. London and Boston, Mass.: Allen & Unwin.

Barth-Eide, W. 1978. Rethinking Food and Nutrition Education Under Changing Socio-Economic Conditions. *Food and Nutrition Bulletin* 2,2: 23–8.

Bates, R. 1981. *Markets and States in Tropical Africa*. Berkeley, Calif.: University of California Press.

—— 1986. Postharvest Considerations in the Food Chain. In: A. Hansen and D. McMillan (eds), *Food in Sub-Saharan Africa*, pp. 239–53. Boulder, Colo.: Lynne Rienner Publishers.

Baulch, B. 1987. Entitlements and the Wollo Famine of 1982–85. *Disasters* 11,3: 195–204.

Baxter, P. 1993. Personal communication with Ian Davis on the evacuation before Pinatubo volcanic eruption and on recovery after Cerro Negro eruption.

Baxter, P.J. and Kapila, M. 1989. Acute Health Impact of the Gas Release at Lake Nyos, Cameroon, 1986. *Journal of Volcanology and Geothermal Research* 39: 266–75.

Bell, B.D., Kara, G., and Batterson, C. 1978. Service Utilization and Adjustment Pattern of Elderly Tornado Victims in an American Disaster. *Mass Emergencies* 3: 71–81.

Benedick, R.E. 1991. *Ozone Diplomacy: New Directions in Safeguarding the Planet*. Cambridge, Mass.: Harvard University Press.

Bennett, O. (ed.) 1991. *Greenwar: Environment and Conflict*. Budapest: Panos Institute.

Berg, A. 1988. Feed the Hungry. *New York Times* 3 September.

Berger, P.L. and Neuhaus, R.J. 1977. *To Empower People; The Role of Mediating Structures in Public Policy*. Washington, DC: American Enterprise Institute for Public Policy Research.

Bernstein, H. 1977. Notes on Capital and Peasantry. *Review of African Political Economy* 10: 60–73.

—— 1990. Taking the Part of Peasants? In: H. Bernstein, B. Crow, M. Mackintosh, and C. Martin (eds), *The Food Question: Profit Versus People?*, pp. 69–79. London: Earthscan.

Berry, S. 1984. The Food Crisis and Agrarian Change in Africa: A Review Essay. *African Studies Review* 27,2: 59–112.

Berz, G. 1990. Natural Disasters and Insurance/Reinsurance. *UNDRO NEWS* January–February: 18–19.

Bie, S. 1990. *Dryland Measurement Techniques*. World Bank, Environment Department Working Paper No. 26. Washington, DC: World Bank.

Biehl, J. 1991. *Rethinking Ecofeminist Politics*. Boston, Mass.: South End Press.

Biswas, M. and Pinstrup-Anderson, P. (eds) 1985. *Nutrition and Development*. London: Oxford University Press.

Black, M. 1992. *A Cause for Our Times – Oxfam the First 50 Years*. (Chapter 6: Acts of God and Acts of Man. pp. 106–31.) Oxford: Oxfam and Oxford University Press.

Blaikie, P.M. 1985a. Natural Resources and the World Economy. In: R.J. Johnston and P.J. Taylor (eds), *A World in Crisis: Geographical Perspectives on Global Problems*, pp. 107–26. London: Blackwell.

—— 1985b. *The Political Economy of Soil Erosion in Developing Countries*. London: Longman.

—— 1989. Explanation and Policy in Land Degradation and Rehabilitation for Developing Countries. *Land Degradation and Rehabilitation* 1,1: 23–38.

Blaikie, P.M. and Brookfield, H. 1987. *Land Degradation and Society*. London: Longman.

Blaikie, P.M., Cameron, J., and Seddon, J.D. 1977. *Centre Periphery and Access in West Central Nepal: Social and Spatial Relations of Inequity*. Monographs in Development Studies, No. 5. University of East Anglia, mimeo, 146pp.

—— 1980. *Nepal in Crisis: Growth and Stagnation at the Periphery*. London and New Delhi: Oxford University Press.

Blaikie, P.M., Harriss, J.C., and Pain, A. 1985. Public Policy and the Utilization of Common Property Resources in Tamil Nadu, India. Report to Overseas Development Administration, Research Scheme R3988.

Blong, R.J. 1984. *Volcanic Hazards, A Sourcebook On the Effects of Eruptions*. New York: Academic Press.

Bohle, H.G., Cannon, T., Hugo, G., and Ibrahim, F.N. (eds) 1991. *Famine and Food Security in Africa and Asia: Indigenous Responses and External Intervention to Avoid Hunger*. Bayreuther Geowissenschaftliche Arbeiten Vol. 15. Bayreuth: Naturwissenschaftliche Gesellschaft Bayreuth.

Bolt, B.A. 1978. *Earthquakes*. San Francisco, Calif.: W.H. Freeman Co.

Bommer, J. 1985. The Politics of Disaster – Nicaragua. *Disasters* 9,4: 270–8.

Bondestam, L. 1974. People and Capitalism in the North Eastern Lowlands of Ethiopia. *Journal of Modern African Studies* 12,3: 423–39.

Borton, J. 1984. *Disaster Preparedness and Response in Botswana*. Report to the Ford Foundation. London: Relief and Development Institute.

—— 1988. Evaluation of ODA Emergency Provision to Africa 1983–86. EV425, August. London: ODA.

Boyce, J. 1987. *Agrarian Impasse in Bengal*. Oxford: Oxford University Press.

—— 1990. Birth of a Megaproject: The Political Economy of Flood Control in Bangladesh. *Environmental Management* 14,4: 419–28.

—— 1992. *Land and Crisis in the Philippines*. London: Macmillan.

Boyden, J. and Davis, I. 1984. Editorial: Getting Mitigation on the Agenda. *Bulletin* 18, October: 2. University of Reading Agricultural Extension and Rural Development Centre.

BRAC (Bangladesh Rural Advancement Committee) 1983a. *The Net: Power Structure in Ten Villages*. Dhaka: BRAC.

BRAC 1983b. *Who Gets What and Why: Resource Allocation in a Bangladesh Village*. Dhaka: BRAC.

Bradley, D. 1977. The Health Implications of Irrigation Schemes and Man-made Lakes in Tropical Environments. In: R. Feachem, M. McGarry, and D. Mara (eds), *Water, Wastes and Health in Hot Climates*, pp. 18–29. Chichester: John Wiley & Sons.

Bradley, P., Raynaut, C., and Torrealba, J. 1977. *The Guidimaka Region of Mauritania: A Critical Analysis Leading to a Development Project*. London: War on Want.

Brammer, H. 1989. Report on the International Conference on the Greenhouse Effect and Coastal Areas of Bangladesh. *Disasters* 13,1: 95.

—— 1990a. Floods in Bangladesh: I. Geographical Background to the 1987 and 1988 Floods. *Geographical Journal* 156,1: 12–22.

—— 1990b. Floods in Bangladesh: II. Flood Mitigation and Environmental Aspects. *Geographical Journal* 156,2: 158–65.

—— 1992. Floods in Bangladesh: Vulnerability and Mitigation Related to Human Settlement. In: Y. Aysan and I. Davis (eds), *Disasters and the Small Dwelling*, pp. 110–18. London: James & James Science Press.

—— 1993. Protecting Bangladesh. *Tiempo: Global Warming and the Third World* 8, April: 7–10.

Brandt, W. 1986. *World Armament and World Hunger*. London: Gollancz.

Branford, S. and Kucinski, B. 1988. *The Debt Squads: The U.S., the Banks, and Latin America*. London: Zed Press.

Bread for the World (ed.) 1991. *Food as a Weapon*. Washington, DC: Bread for the World.

Brennan, T. 1987. *Uprooted Angolans: From Crisis to Catastrophe*. Washington, DC: US Committee for Refugees.

Briggs, P. 1973. *Rampage: The Story of Disastrous Floods, Broken Dams, and Human Fallibility*. New York: David McKay.

British Overseas Development 1990. Holding Back the Flood: Action Planned to Help Save Bangladesh. *British Overseas Development* 10, February: 1, 4.

Broad, W.J. 1992. Breaking a Date with Doomsday. *New York Times* 1 April: A16.

Brokensha, D., Warren, D., and Werner, O. (eds) 1980 *Indigenous Knowledge Systems and Development*. Lanham, Md.: University Press of America.

Brown, E.P. 1991. Sex and Starvation: Famine and Three Chadian Societies. In: R. Downs, D. Kerner, and S. Reyna (eds), *The Political Economy of African Famine*, pp. 293–321. Philadelphia: Gordon & Breach Science Publishers.

Brownlea, A.A. 1981. From Public Health to Political Epidemiology. *Social Science and Medicine* 15D: 57–67.

Brush, L.M, Wolman, M.G., and Bing-Wei, H. 1989. *Taming the Yellow River: Silt and Floods*. Dordrecht: Kluwer.

Brush, S.B. 1987. Diversity and Change in Andean Agriculture. In: P. Little and M. Horowitz (eds), *Lands at Risk in the Third World*, pp. 271–89. Boulder, Colo.: Westview.

Bryant, E.A. 1991. *Natural Hazards*. Cambridge: Cambridge University Press.

Bryceson, D.F. 1989. Nutrition and the Commoditization of Food in Sub-Saharan Africa. *Social Science and Medicine* 28,5: 425–40.

BSSA 1981–9. Seismological Notes. *Bulletin of Seismological Society of America* 71–9.

Burton, I. and Kates, R.W. 1964. Perception of Natural Hazards in Resource Management. *Natural Resources Journal* 3: 412–41.

Burton, I., Kates, R.W., and White, G.F. 1978. *The Environment as Hazard*. New York: Oxford University Press.

Bush, R. 1985. Drought and Famines. *Review of African Political Economy* 33: 59–63.

Button, G. 1992. When Marsians Take Over: The Politics of Symbolic Resistance in Mars Cove, Alaska. Paper presented at the 51st Annual Meeting of the Society for Applied Anthropology, Memphis, Tennessee.

Byrne, L. 1988. Tree Felling Blamed for Rio Disaster. *Observer* (London) 28 February: 19.

Cain, M. 1978. The Household Lifecycle and Economic Mobility in Bangladesh. Centre for Policy Studies Working Paper, Population Council, New York.

Cairncross, S. 1988. Domestic Water Supply in Rural Africa. In: D. Rimmer (ed.), *Rural Transformation in Tropical Africa*, pp. 46–65. Athens, OH: Ohio University Press.

Cairncross, S., Hardoy, J., and Satterthwaite, D. 1990a. New Partnerships for Healthy Cities. In: S. Cairncross, J. Hardoy, and D. Satterthwaite (eds), *The Poor Die Young: Housing and Health in Third World Cities*, pp. 245–68. London: Earthscan.

—— (eds) 1990b. *The Poor Die Young: Housing and Health in Third World Cities*. London: Earthscan.

Caldwell, J.C., Reddy, P.H., and Caldwell, P. 1986. Period High Risk as a Cause of Fertility Decline in a Changing Rural Environment: Survival Strategies in the 1980–1983 South Indian Drought. *Economic Development and Cultural Change* 34.

Campbell, D. 1987. Participation of a Community in Social Science Research: A Case Study from Kenyan Maasailand. *Human Organization* 46,2: 160–7.

Campbell, J.R. 1984. *Dealing with Disaster, Hurricane Responses in Fiji*. Suva, Fiji: Pacific Islands Development Programme, East–West Center, and the Government of Fiji.

Cannon, T. 1991. Hunger and Famine: Using a Food Systems Model to Analyse Vulnerability. In: H.G. Bohle, T. Cannon, G. Hugo, and F.N. Ibrahim (eds), *Famine and Food Scarcity in Africa and Asia: Indigenous Responses and External Intervention to Avoid Hunger*, pp. 291–312. Bayreuther Geowissenschaftliche Arbeiten Vol.15. Bayreuth: Naturwissenschaftliche Gesellschaft Bayreuth.

—— 1994. Vulnerability Analysis and the Explanation of 'Natural' Disasters. In: A. Varley (ed.), *Disasters, Development and the Environment*. London: Belhaven Press.

Cardona, O.D. and Sarmiento, J.P. 1990. *Vulnerability Analysis and Risk Assessment for the Health of a Community Exposed to Disasters*. Bogota: Colombian Red Cross.

Carlstein, T. 1982. *Time Resources, Society and Ecology*. London: George Allen & Unwin.

Carney, J. 1988. Struggles Over Crop Rights and Labour Within Contract Farming Households in a Gambian Irrigated Rice Project. *Journal of Peasant Studies* 15,3: 334–49.

Carson, R. 1962. *Silent Spring*. Harmondsworth: Penguin.

Carter, R.W.G. 1987. Man's Response to Sea-Level Change. In: R.J.N. Devoy (ed.), *Sea Surface Studies*, pp. 464–98. London: Croom Helm.

Castro, F. 1984. *The World Crisis: Its Economic and Social Impact on the Underdeveloped Countries*. Morant Bay, Jamaica, London, and Haarlem: Maroon Publishing House, Zed Press, and In De Knipscheer.

Castro, J. 1989. The Benefits of Being Prepared. *Time* 44, 30 October.

Cater, N. 1986. *Sudan: The Roots of Famine*. Oxford: Oxfam.

Cedeno, J.E.M. 1986. Rainfall and Flooding in the Guayas River Basin and Its Effects on the Incidence of Malaria 1982–1985. *Disasters* 10,2: 107–11.

Chairetakis, A. 1992. Past as Present: History and Reconstruction after the 1980 Earthquake in Campania, Southern Italy. Paper presented at the 51st Annual Meeting of the Society for Applied Anthropology, Memphis, Tennessee.

Chambers, R. 1983. *Rural Development: Putting the Last First*. New York: Longman.

—— 1989. Editorial Introduction: Vulnerability, Coping and Policy. *IDS Bulletin* 20,2: 1–7

Chambers, R., Longhurst, R., and Pacey, A. (eds) 1981. *Seasonal Dimensions to Rural Poverty*. London: Francis Pinter.

Chambers, R., Pacey, A., and Thrupp, L. (eds) 1989. *Farmer First*. London: Intermediate Technology Publications.

Chambers, R., Saxena, N., and Shah, T. 1990. *To the Hands of the Poor: Water and Trees*. Boulder, Colo.: Westview.

Chapin, G. and Wasserstrom, R. 1981. Agricultural Production and Malaria Resurgence in Central America and India. *Nature* 293, 5829, 17 September: 181–5.

Chaturvedi, M.C. 1981. Flood Management – New Concepts, Technology and Planning Approach. Paper presented at International Conference on Flood Disaster, New Delhi.

Chen, L. (ed.) 1973. *Disaster in Bangladesh*. Oxford: Oxford University Press.

Cheru, F. 1989. *The Silent Revolution in Africa: Debt, Development and Democracy*. Harare and London: Anvil Press and Zed Press.

Chirimuuta, R.C. and Chirimuuta, R.J. 1987. *Aids, Africa and Racism*. Bretby, Derbyshire: privately published.

Chowdhury, J.U. 1991. Flood Action Plan: One Sided Approach? *Bangladesh Environmental Newsletter* 2,2: 1, 3. Dhaka: Bangladesh Centre for Advanced Studies.

Christodoulou, D. 1990. *The Unpromised Land: Agrarian Reform and Conflict Worldwide*. London: Zed Press.

Chung, J. 1987. Fiji, Land of Tropical Cyclones and Hurricanes: A Case Study of Agricultural Rehabilitation. *Disasters* 11,1: 40–8.

CIIR (Catholic Institute of International Relations) 1975. *Honduras: Anatomy of a Disaster*. London: CIIR.

CIMADE, INODEP, and MINK 1986. *Africa's Refugee Crisis*. London: Zed Press.

Clark, C. 1982. *Flood*. Alexandria, Va: Time-Life Books.

Clark, J. 1991. *Democratizing Development: The Role of Voluntary Organizations*. West Hartford, Conn.: Kumarian Press.

Clark, W. 1989. Managing Planet Earth. *Scientific American* 262,3: 46–57.

Clarke, J.I. (ed.) 1989. *Population and Disaster*. Oxford: Blackwell.

Clay, E. 1985. The 1974–1984 Floods in Bangladesh: From Famine to Food Crisis Management. *Food Policy* 10,3: 202–6.

Clay, J. 1988. *Indigenous Peoples and Tropical Forests: Models of Land Use and Management from Latin America*. Cambridge, Mass.: Cultural Survival.

Clay, J. and Holcomb, B. 1985. *Politics and the Ethiopian Famine 1984–1985*. Cambridge, Mass.: Cultural Survival.

Clay, J., Steingraber, S., and Niggli, P. 1988. *The Spoils of Famine: Ethiopian Famine Policy and Peasant Agriculture*. Cambridge, Mass.: Survival International.

Cliff, A.D. and Smallman-Raynor, M.R. 1992. The Aids Pandemic: Global Geographical Patterns and Local Spatial Processes. *Geographical Journal* 158, 2, July: 182–98.

Cliff, J. 1991. The War on Women in Mozambique: Health Consequences of South African Destabilization, Economic Crisis, and Structural Adjustment. In: M. Turshen (ed.), *Women and Health in Africa*, pp. 15–34. Trenton, NJ: Africa World Press.

Cliffe, L. and Moorsom, R. 1979. Rural Class Formation and Ecological Collapse in Botswana. *Review of African Political Economy* 15–16: 35–52.

Coburn, A.W., Hughes, R.E., Illi, D., Nash, D.F.T, and Spence, R.J.S. 1984. The Construction and Vulnerability to Earthquakes of Some Building Types in Northern Areas of Pakistan. In: K.J. Miller (ed.), *The International Karakoram Project*, vol. 2, pp. 228–37. Cambridge: Cambridge University Press.

Coburn, A.W. and Spence, R.J.S. 1992. *Earthquake Protection*. Chichester: Wiley.

Coburn, A., Spence, R.W., and Pomonis. 1991. *Vulnerability and Risk Assessment*. Trainers and Trainees Guide. Disaster Management Training Programme (DMTP), Geneva: UNDRO/UNDP.

Cochrane, H.C. 1975. *Natural Hazards and Their Distributive Effects*. National Science Foundation Program on Technology, Environment and Man Monograph NSF-RA-E-75-003. Boulder, Colo.: University of Colorado, Institute of Behavioral Science.

Cohen, D. 1991. *Aftershock – The Psychological and Political Consequences of Disaster*. London: Paladin Grafton Books.

Cohen, J.M. and Lewis, D.B. 1987. Role of Government in Combatting Food Shortages: Lessons from Kenya 1984–85. In: M. Glantz (ed.), *Drought and Hunger in Africa*, pp. 269–96. Cambridge: Cambridge University Press.

Cohen, M.M. 1977. *The Food Crisis in Prehistory: Overpopulation and the Origins of Famine*. New Haven, Conn.: Yale University Press.

Cohen, S.P. and Raghavulu, C.V. 1979. *The Andhra Cyclone of 1977*. New Delhi: Vikas.

Collins, J. 1989. *Fire on the Rim*. San Francisco, Calif.: Food First.

Conroy, C. and Litvinoff, M. 1988. *The Greening of Aid*. London: Earthscan.

Conway, G.R. and Barbier, E.B. 1990. *After the Green Revolution: Sustainable Agriculture for Development*. London: Earthscan.

Cook, H.L. and White, G.F. 1962. Making Wise Use of Flood Plains. In: Anon. (ed.), *United Nations Conference on Applications of Science and Technology*, vol. 1, pp. 343–59. Washington, DC: Government Printing Office.

Copans, J. (ed.) 1975. *Secheresses et famines du Sahel*, 2 vols. Paris: Maspero.

—— 1983. The Sahelian Drought: Social Sciences and the Political Economy of Underdevelopment. In: K. Hewitt (ed.), *Interpretations of Calamity*, pp. 83–97. Boston, Mass.: Allen & Unwin.

Corbett, J. 1988. Famine and Household Coping Strategies. *World Development* 16: 1099–1112.

Cornia, G., Jolly, R., and Stewart, F. (eds) 1987. *Adjustment with a Human Face*, 2 vols. New York: Oxford University Press.

Coulson, A. 1982. *The Political Economy of Tanzania*. Oxford: Oxford University Press.

Cowie, L. 1972. *The Black Death and Peasants Revolt*. London: Wayland.

Crosby, A. 1986. *Ecological Imperialism: The Biological Expansion of Europe, 900–1900*. Cambridge: Cambridge University Press.

—— 1991. The Biological Consequences of 1492. *Report on the Americas* 25,2: 6–13.

Crossette, B. 1992. Sudan is Said to Force 400,000 People Into Desert. *New York Times* 22 February: 5.

Crow, B. 1984. Warnings of Famine in Bangladesh. *Economic and Political Weekly* 19,40: 1754–8.

—— 1990. Moving the Lever: A New Food Aid Imperialism? In: H. Bernstein, B. Crow, M. Mackintosh, and C. Martin (eds), *The Food Question: Profits Versus People*, pp. 32–42. London: Earthscan.

Cuenya, B., Almada, H., Armus, H., Castells, J., di Loreto, M., and Penalva, S. 1990. Community Action to Address Housing and Health Problems: The Case of San Martin in Buenos Aires, Argentina. In: S. Cairncrossm J. Hardoy, and D. Satterthwaite (eds), *The Poor Die Young: Housing and Health in Third World Cities*, pp. 25–55. London: Earthscan.

Cuny, F.C. 1983. *Disasters and Development*. New York: Oxfam and Oxford University Press.

—— 1987. Sheltering the Urban Poor, Lessons and Strategies of the Mexico City and San Salvador Earthquakes. *Open House International* 12,3: 16–20.

Currey, B. 1978. The Famine Syndrome: Its Definition for Preparedness and Prevention in Bangladesh. *Ecology of Food and Nutrition* 7,1.

—— 1981. The Famine Syndrome: Its Definition for Relief and Rehabilitation in Bangladesh. In: J.R.K. Robson (ed.), *Famine: Its Causes, Effects and Management*. New York: Gordon & Breach.

—— 1984. Coping with Complexity in Food Crisis Management. In: B. Currey and G. Hugo (eds), *Famine as a Geographical Phenomenon*, pp. 183–202. Dordrecht: D. Reidel.

Curson, P. 1989. Introduction. In J.I. Clarke (ed.), *Population and Disaster*, pp. 1–23. Oxford: Blackwell.

Curtis, D., Hubbard, M., and Shepherd, A. (eds) 1988. *Preventing Famine: Policies and Prospects for Africa*. London: Routledge.

Cutler, P. 1984. Famine Forecasting: Prices and Peasant Behaviour in Northern Ethiopia. *Disasters* 8,1: 48–55.

—— 1985. Detecting Food Emergencies: Lessons from the Bangladesh Crisis. *Food Policy* 10.

—— 1986. The Response to Drought of Beja Famine Refugees in Sudan. *Disasters* 9.

Dahl, G. and Hjort, G. 1976. *Having Herds: Pastoral Herd Growth and Household Economy*. Stockholm Studies in Social Anthropology 2. Stockholm: Department of Social Anthropology, University of Stockholm.

Dalal-Clayton, B. 1990. *Environmental Aspects of the Bangladesh Flood Action Plan*. Gatekeeper Series No. 1. London: International Institute for Environment and Development (IIED).

Dando, W. 1980. *The Geography of Famine*. London: Arnold.

—— 1981. Man-Made Famines: Some Geographical Insights from an Exploratory Study of a Millenium of Russian Famines. In: J. Robson (ed.), *Famine: Its Causes, Effects and Management*. New York: Gordon & Breach.

Dankelman, I. and Davidson, J. 1988. *Women and Environment in the Third World*. London: Earthscan.

Davenport, W. 1960. *Jamaican Fishing: A Game Theory Analysis*. Yale University Publications in Anthropology 59. New Haven, Conn.: Yale University Press.

Davis, I. 1977a. Emergency Shelter. *Disasters* 1, 1: 23–40.

—— 1977b. The Intervenors. *New Internationalist* 53: 21–3.

—— 1978. *Shelter After Disaster*. Oxford: Oxford Polytechnic Press.

—— (ed.) 1981. *Disasters and the Small Dwelling*. Oxford: Pergamon Press.

—— 1984a. A Critical Review of the Work Method and Findings of the Housing and Natural Hazards Group. In: K.J. Miller (ed.), *The International Karakoram Project*, vol. 2, pp. 200–27. Cambridge: Cambridge University Press.

—— 1984b. The Squatters Who Live Next Door to Disaster. *Guardian* (London) 7 December: 7.

—— 1986. The Planning and Maintenance of Urban Settlements to Resist Extreme Climatic Forces. In: T.R. Oke (ed.), *Urban Climatology and its Applications with Special Regard to Tropical Areas*, pp. 277–312. World Climate Programme – Proceedings of the Technical Conference, Mexico. Geneva: World Meteorological Organization.

—— 1987. Safe Shelter Within Unsafe Cities: Disaster Vulnerability and Rapid Urbanisation. *Open House International* 12,3: 5–15.

—— 1988. Acts of God Increasingly Amount to Acts of Criminal Negligence. *Guardian* (London) 30 December: 7.

Davis, I. and Bickmore, D. 1993. Data Management for Disaster Planning. In: The Royal Society, Proceedings of Conference: Natural Disasters – Protecting Vulnerable Communities (13–15 October 1993). London: Royal Society, Royal Academy of Engineering, and the Society for Earthquake and Civil Engineering Dynamics (SECED).

Davis, I. and Gupta, S.P. 1991. Technical Background Paper. In: Asian Development Bank, *Disaster Mitigation in Asia and the Pacific*, pp. 23–69. Manila: Asian Development Bank.

Davis, I., Kishigami, H., Takei, S., Yaoxian, Y., and Johansson, M. 1992. Rehabilitation Assistance to Anhui Province Following the Flood Disaster, May–July 1991. Report of UNDP Appraisal Mission, 4–16 December 1991. CPR/91/712. Beijing: UNDP.

Davis, S.H. and Hodson, J. 1982. *Witnesses to Political Violence in Guatemala. The Suppression of a Rural Development Movement*. Boston, Mass.: Oxfam America.

De Beer, C. 1986. *The South African Disease*. Trenton, NJ: Africa World Press.

De Castro, J. 1957. *Le livre noir de la faim*. Paris: Editions Ouvrières.

—— 1966. *Death in the Northeast*. New York: Vintage.

—— 1977. *Geopolitics of Hunger*. New York: Monthly Review. (First published as *Géopolitique de la faim*. Paris: Editions Ouvrières 1952.)

de Milan, C. 1988. Personal communication with I. Davis in Ibagué, Colombia.

De Ville, C. and Lechat, M. 1976. Health Aspects in Natural Disasters. *Tropical Doctor* October: 168–70.

De Vylder, S. 1982. *Agriculture in Chains: Bangladesh: A Case Study in Contradictions and Constraints*. London: Zed Press.

de Waal, A. 1987. The Perception of Poverty and Famines. *International Journal of Moral and Social Studies* 2,3.

—— 1989a. Famine Mortality: A Case Study of Darfur, Sudan 1984–5. *Population Studies* 43,1.

—— 1989b. *Famine That Kills. Darfur, Sudan, 1984–1985*. Oxford: Clarendon Press.

—— 1991. Famine and Human Rights. *Development in Practice: An Oxfam Journal* 1,2: 77–83.

de Waal, A. and Amin, M.M. 1986. *Report on Save the Children Fund Activities in Darfur*. Nyala, Sudan: Save the Children Fund.

Debach, P. 1974. *Biological Control by Natural Enemies*. Cambridge: Cambridge University Press.

249

Degg, M.R. 1989. Earthquake Hazard Assessment after Mexico 1985. *Disasters* 13,3: 237–54.

—— 1992. Reducing Vulnerability to Earthquake Hazard in the Third World: Recent Initiatives by the International Reinsurance Market. In: A. Varley (ed.), *Disasters, Development and the Environment*. London: Belhaven Press.

Deming, A. and Theodore, S. 1989. The Earthquake that Created Italian Gold. *Newsweek* 9 January: 13.

Deny, F.D. and Minear, L. 1992. *The Challenges of Famine Relief Emergency Operations in the Sudan*. Washington, DC: Brookings Institution.

Devereux, S. 1987. FAO & FED = Famine: Not a Refutation of Professor Sen's Theory. Paper at Workshop 'The Causes of Famine', Queen Elizabeth House, Oxford, 9 May.

Devereux, S. and Hay, R. 1986. *Origins of Famine. A Review of the Literature*. Oxford: University of Oxford, Food Studies Group.

Dey, J. 1981. Gambian Women: Unequal Partners in Rice Development Projects? *Journal of Development Studies* 17,3: 109–22.

Dimbleby, J. 1983. *The Unknown Famine*. BBC TV.

Diriba, K. 1991. Famines and Food Security in Kembatana Hadiya, Ethiopia. A Study in Household Survival Strategies. Ph.D. Thesis, University of East Anglia.

Disaster Preparedness Bureau 1988. Information provided by the Disaster Preparedness Bureau, National Land Agency, Prime Minister's Office, Government of Japan. May.

Dixon, J., Carpenter, R., Fallon, L., Sherman, P., and Manipomoke, S. 1988. *Economic Analysis of the Environmental Impacts of Development Projects*. London: Earthscan.

Doerner, W.R. 1985. Last Rites for a Barrio: A Crushing Mud Slide Kills Hundreds in Puerto Rico. *Time* 21 October: 32.

Donohue, J. 1982. Some Facts and Figures on Urbanisation in the Developing World, *Assignment Children* 57,8.

Doughty, P.L. 1970. 'What Will We do When the Rains Come?' Unpublished report, Peru Earthquake Committee. (Cited in A. Oliver-Smith 1986b. *The Martyred City: Death and Rebirth in the Andes*. Albuquerque, NM: University of New Mexico Press.)

Douglas, M. 1985. *Risk Acceptability According to the Social Sciences*. London and New York: Routledge and Russell Sage Foundation.

Douglas, M. and Wildavsky, A. 1982. *Risk and Culture: An Essay on the Selection of Technical and Environmental Dangers*. Berkeley, Calif.: University of California Press.

Downing, T.E. 1991. *Assessing Socioeconomic Vulnerability to Famine: Frameworks, Concepts and Applications*. Alan Shawn Feinstein World Hunger Programme. Providence, RI: Brown University.

—— 1992. *Climate Change and Vulnerable Places: Global Food Security and Country Studies in Zimbabwe, Kenya, Senegal and Chile*. Environmental Change Unit (ECU) Research Report No. 1. Oxford: University of Oxford.

Downing, T., Gitu, K., and Kamau, C. (eds) 1989. *Coping with Drought in Kenya: National and Local Strategies*. Boulder, Colo.: Lynne Rienner.

Downs, R.E., Kerner, D.O., and Reyna, S.P. (eds) 1991. *The Political Economy of African Famine*. Philadelphia: Gordon & Breach Science Publishers.

Dowrick, D.J. 1977. *Earthquake Resistant Design*. Chichester: Wiley.

Doyal, L. 1981. *The Political Economy of Health*. London: Pluto Press.

Doyal, L. and Gough, I. 1991. *A Theory of Human Need*. London and New York: Macmillan and Guilford Press.

Drabek, T. 1986. *Human Systems Response to Disaster*. London: Pergamon.

Drèze, J. 1988. *Famine Prevention in India*. Development Economics Research Programme. No.3, January. London: London School of Economics.

Drèze, J. and Sen, A. 1989. *Hunger and Public Policy*. Oxford: Clarendon Press.

—— (eds) 1990a. *The Political Economy of Hunger*. Vol. 1: *Entitlement and Well Being*. Oxford: Clarendon Press.

—— (eds) 1990b. *The Political Economy of Hunger*. Vol. 2: *Famine Prevention*. Oxford: Clarendon Press.

—— (eds) 1990c. *The Political Economy of Hunger*. Vol. 3: *Endemic Hunger*. Oxford: Clarendon Press.

D'Souza, F. 1984. The Socio-Economic Cost of Planning for Hazards. An Analysis of Barculti Village, Yasin, Northern Pakistan. In: K.J. Miller (ed.), *The International Karakoram Project*, vol. 2, pp. 289–322. Cambridge: Cambridge University Press.

—— 1988. Famine: Social Security and an Analysis of Vulnerability. In: G.A. Harrison (ed.), *Famine*, pp. 1–56. Oxford: Oxford University Press.

DSWD (n.d.) (*c.* 1989). *Proceedings of the First National Disaster Management Workshop*, 6–18 March. Department of Social Welfare and Development (DSWD) and NDCC Inter-Agency Planning Committee. Tagaytay City: Development Academy of the Philippines. (Simulation exercise on local level co-ordination – Taal volcano, pp. 100–6.)

Dudley, E. 1988. Disaster Housing: Strong Houses or Strong Institutions? *Disasters* 12,2: 111–21.

Dunham, A. 1959. Flood Control Via the Police Power. *University of Pennsylvania Law Review* 107: 1098–132.

During, A. 1989. Mobilizing at the Grassroots. In: L. Brown *et al.* (eds), *State of the World 1989*, pp. 154–73. New York: Norton.

Dynes, R.R., De Marchi, B., and Pelanda, C. (eds) 1987. *Sociology of Disaster*. Milan: Franco Agneli Libri.

Eagleman, Joe R., 1983. *Severe and Unusual Weather*. New York: Van Nostrand Reinhold.

Ebert, Charles H.V. 1988. *Disasters: Violence of Nature and Threats by Man*. Dubuque, Ia.: Kendall/Hunt.

Echeverria, E. 1991. Decentralising Mexico's Health Care Facilities. In: A. Kreimer and M. Munasinghe (eds), *Managing Natural Disasters and the Environment*, pp. 60–1. Washington, DC: World Bank.

Eckholm, E. 1976. *Losing Ground*. Oxford: Pergamon.

Economist, The 1989. Score One For the Trees. 14 January: 53.

Ehrlich, P. and Ehrlich, A. 1990. *The Population Explosion*. New York: Simon & Schuster.

Eide, A., Eide, W.B., Goonatilake, S., Gussow, J., and Omawale (eds) 1984. *Food as a Human Right*. Tokyo: United Nations University Press.

Ekejuiba, F. 1984. Contemporary Households and Major Socio-Economic Transitions in Eastern Nigeria: Toward a Reconceptualisation of the Household. In: J.I. Guyer and P.E. Peters (eds), *Conceptualising in the Household*, pp. 9–13. Cambridge, Mass.: Harvard University Press.

Elahi, K.M. 1989. Population Displacement Due to Riverbank Erosion of the Jamuna in Bangladesh. In: J.I. Clarke (ed.), *Population and Disaster*, pp. 81–97. Oxford: Basil Blackwell.

Ellis, F. 1988. *Peasant Economics: Farm Households and Agrarian Development*. Cambridge: Cambridge University Press.

Emel, J. and Peet, R. 1989. Resource Management and Natural Hazards. In: R. Peet and N. Thrift (eds), *New Models in Geography*. vol. 1, pp. 49–76. London: Unwin Hyman.

Ennew, J. and Milne, B. 1989. *The Next Generation: Lives of Third World Children*. London: Zed Press.

EPOCA (The Environmental Project on Central America) 1990. *Guatemala: A Political Ecology*. Green Paper No. 5. San Francisco, Calif.: Earth Island Institute.

Farooque, M. 1993. A Legal Perspective on the FAP. *Tiempo: Global Warming and the Third World* 8, April: 17–19.

Feachem, R., McGarry, M., and Mara, D. (eds) 1978. *Water, Wastes and Health in Hot Climates*. Chichester: John Wiley & Sons.

Feierman, S. 1985. Struggles for Control: The Social Roots of Health and Healing in Modern Africa. *African Studies Review* 28,2–3: 73–147.

Fernandes, W. and Menon, G. 1987. *Tribal Women and Forest Economy: Deforestation, Exploitation and Status Change*. New Delhi: Indian Social Institute.

Fernando, A. 1990. The Role of Nongovernmental Organizations in Sri Lanka. In: A. Kreimer and M. Munasinghe (eds), *Managing Natural Disasters and the Environment*, pp. 172–81. Washington, DC: World Bank.

Firth, R. 1959. *Social Change in Tikopia*. London: Allen & Unwin.

Fiselier, J.L. 1990. *Living Off the Floods: Strategies for the Integration of Conservation and Sustainable Resource Utilization in Floodplains*. Leiden: Environmental Database on Wetland Interventions.

Food and Agriculture Organization (FAO) 1982. *Potential Population Supporting Capacities of Lands in the Developing World*. Technical Report of Project FPA/INT/513. Rome: FAO/UNFPA/IIASA.

Ford, K. 1987. Private correspondence with Ian Davis.

Forde, D. 1972. *Trypanosomiasis in Africa*. Oxford: Oxford University Press.

Forse, B. 1989. The Myth of the Marching Desert. *New Scientist* 4 February: 31–2.

Foster, H. 1980. *Disaster Planning: The Preservation of Life and Property*. New York: Springer-Verlag.

Fowler, C. and Mooney, P. 1990. *Shattering: Food, Politics, and the Loss of Genetic Diversity*. Tucson, Ariz.: University of Arizona Press.

Franke, R. 1984. Tuareg of West Africa. In: D. Stea and B. Wisner (eds), *The Fourth World: The Geography of Indigenous Struggles*. Thematic issue of *Antipode* 16,2: 45–53.

Franke, R. and Chasin, B.H. 1980. *Seeds of Famine: Ecological Destruction and the Development Dilemma in the Western Sahel*. Montclair, NJ: Allenheld, Osmun.

—— 1989. *Kerala: Radical Reform as Development in an Indian State*. San Francisco, Calif.: Institute for Food and Development Policy.

Frazier, K. 1979. *The Violent Face of Nature*. New York: William Morrow.

Freeberne, M. 1993. A Reconstruction of the 1991 Floods in China – Natural or Man-Made Disasters? *Disaster Management* 5,2: 67–79.

French, R.A. 1989. Houses Built on Sand. *Geographical Magazine* March: 32–4.

Gamser, M.S. 1988. *Power from the People: Innovation, User Participation, and Forest Energy Development*. London: IT Publications.

Garnsey, P. 1988. *Famine and Food Supply in the Graeco-Roman World: Responses to Risks and Crises*. Cambridge: Cambridge University Press.

George, S. 1988. *A Fate Worse Than Debt: The Third World Financial Crisis and the Poor*. London: Penguin.

Giesecke, A. 1983. Case History of the Peru Prediction for 1980–81. In: *Proceedings of the Seminar on Earthquake Case Histories*. Geneva: UNDRO, pp. 51–75.

Gill, P. 1986. *A Year in the Death of Africa: Politics, Bureaucracy and the Famine*. London: Paladin.

Gini, C. and De Castro, J. (eds) 1928. *Materiaux pour l'étude des calamités*. Geneva: League of Nations.

Glantz, M. (ed.) 1987. *Drought and Hunger in Africa*. Cambridge: Cambridge University Press.

Goheen, M. 1991. Ideology, Gender and Change: Social Relations of Production and Reproduction in Nso, Cameroon. In: R. Downs, D. Kerner, and S. Reyna (eds), *The Political Economy of African Famine*, pp. 273–92. Philadelphia: Gordon & Breach Science Publishers.

Goldstein, G. 1990. Life Saving Services. In: S. Cairncross, J. Hardoy, and D. Satterthwaite (eds), *The Poor Die Young: Housing and Health in Third World Cities*, pp. 213–27. London: Earthscan.

Gomez, M.A. 1991. Reducing Urban and Natural Risks in Mexico City. In: A. Kreimer and M. Munasinghe (eds), *Managing Natural Disasters and the Environment*, pp. 56–7. Washington, DC: World Bank.

Goodfield, J. 1991. *The Planned Miracle*. London: Cardinal Books.

Gould, P. 1969. Man Against His Environment: A Game Theoretic Framework. In: A. Vayda (ed.), *Environment and Cultural Behavior*, pp. 234–51. Garden City, NY: The Natural History Press.

Government of India 1978. Report of the Working Group on Integrated Action Plan for Flood Control (in Indo-Gangetic Basin). New Delhi: Ministry of Agriculture and Irrigation (Department of Agriculture).

Goyder, H. and Goyder, C. 1988. Case Studies of Famine: Ethiopia. In: D. Curtis, M. Hubbard, and A. Shepherd (eds), *Preventing Famine: Policies and Prospects for Africa*, pp. 73–110. London: Routledge.

Graff, T.O. and Wiseman, R.F. 1978. Changing Concentrations of Older Americans. *Geographical Review* 68: 379–93.

Grainger, A. 1990. *The Threatening Desert: Controlling Desertification*. London: Earthscan.

Grainger, O.E. 1990. *Natural Disasters and Social Change: An Eastern Caribbean Perspective*. Berkeley, Calif.: privately published.

Green, C. 1990. Personal communication with Terry Cannon.

Green, D. 1988. Nicaragua's Double Hurricane. *Guardian* (London) 3 December.

Green, M. 1979. Today's Children – Tomorrow's Parents. *Agenda* 2,1: 1–5.

Greenough, P.R. 1982. *Prosperity and Misery in Modern Bengal: The Famine of 1943–44*. Oxford: Oxford University Press.

Griggs, G.B. and Gilchrist, J.A. 1983. *Geological Hazards, Resources, and Environmental Planning*, 2nd edn. Belmont, Calif.: Wadsworth.

Guardian 1981. Predicted Earthquake in Lima. 25 June.

—— 1991. Waiting for the Wave of Death, Special Report on African Famine Risk. 26 April: 24.

Gueri, M., Gonzalez, C., and Morin, V. 1986. The Effect of the Floods Caused by El Nino on Health. *Disasters* 10,2: 118–24.

Guillette, E.A. 1991. The Impact of Recurrent Disaster on the Aged of Botswana. Paper presented at the 50th Annual Meeting of the Society for Applied Anthropology, Charleston, South Carolina.

—— 1992. Leading and Following During Disaster: An Age-Group Approach to the Recovery Process. Paper presented at the 51st Annual Meeting of the Society for Applied Anthropology, Memphis, Tennessee.

Gupta, A. 1988. *Ecology and Development in the Third World*. London: Routledge.

Gurdilek, R. 1988. Sniffer Dogs Search for Landslide Victims as Hopes Fade. *The Times* 25 June: 7.

Guyer, J. 1981. The Household in African Studies. *African Studies Review* 24,2–3: 87–137.

Guyer, J.I. and Peters, P.E. (eds) 1984. *Conceptualising the Household*. Cambridge, Mass.: Harvard University Press.

Hagman, G. 1984. *Prevention Better than Cure: A Swedish Red Cross Report on Human and Environmental Disasters in the Third World*. Stockholm: Swedish Red Cross.

Hancock, G. 1989. *The Lords of Poverty*. London: Macmillan.

Handmer, J. and Penning-Rowsell, E. (eds) 1990. *Hazards and the Communication of Risk*. Aldershot: Gower Publishing.

Hansen, A. and McMillan, D.E. (eds) 1986. *Food in Sub-Saharan Africa*. Boulder, Colo.: Lynne Rienner.

Hansen, A. and Oliver-Smith, A. (eds) 1982. *Involuntary Migration and Resettlement: The Problems and Responses of Dislocated People*. Boulder, Colo.: Westview.

Hansen, E. (ed.) 1987. *Africa: Perspectives on Peace and Development*. London: Zed Press.

Hansen, S. 1988. Structural Adjustment Programs and Sustainable Development. Paper commissioned by UNEP for the Annual Session of the Committee of International Development Institutions on the Environment (CIDIE), 13–17 June. Washington, DC.

Hanson, W.J. 1967. *East Pakistan in the Wake of the Cyclone*. London: Longman.

Haq, K. and Kirdar, U. (eds) 1987. *Human Development, Adjustment and Growth*. Islamabad: North South Roundtable.

Haque, C.E. 1988. Impact of River-Bank Erosion Hazard in the Brahmaputra–Jamuna Floodplain: A Study of Population Displacement and Response Strategies. Ph.D. Dissertation, University of Manitoba.

Haque, C.E. and Blair, D. 1992. Vulnerability to Tropical Cyclones: Evidence from the April 1991 Cyclone to Coastal Bangladesh. *Disasters* 16,3: 217–29.

Hardoy, J. 1987. Natural Disasters in Latin America. Unpublished Report for the Institute of the Environment and Development. London: IIED.

Hardoy, J.E. and Satterthwaite, D. 1989. *Squatter Citizen: Life in the Urban Third World*. London: Earthscan.

Harley, R.M. 1990. *Breakthroughs on Hunger*. Washington, DC: Smithsonian Institute Books.

Harrell-Bond, B. 1986. *Imposing Aid: Emergency Assistance to Refugees*. Oxford: Oxford University Press.

Harrison, P. 1987. *The Greening of Africa*. London: Penguin.

Harrison, P. and Palmer, R. 1986. *News out of Africa: Biafra to Band Aid*. London: Hilary Shipman.

Harriss, B. 1988. Limitations of the Lessons from India. In: D. Curtis, M. Hubbard, and A. Shepherd (eds), *Preventing Famine. Policies and Prospects for Africa*, pp. 157–61. London: Routledge.

Hart, J.T. 1971. The Inverse Care Law. *Lancet* i: 405–12.

Hartmann, B. 1987. *Reproductive Rights and Wrongs*. New York: Harper & Row.

Hartmann, B. and Boyce, J. 1983. *A Quiet Violence: View from a Bangladesh Village*. London: Zed Press.

Hartmann, B. and Standing, H. 1989. *The Poverty of Population Control: Family Planning and Health Policy in Bangladesh*. London: Bangladesh International Action Group.

254

Havlick, S.W. 1986. Third World Cities at Risk: Building for Calamity. *Environment* 28,9, November: 6.

Hayter, T. and Watson, C. 1985. *Aid: Rhetoric and Reality*. London: Pluto Press.

Hecht, S. and Cockburn, A. 1989. *The Fate of the Forest: Developers, Destroyers and Defenders of the Amazon*. London: Verso.

Hellden, U. 1984. Drought Impact Monitoring: A Remote Sensing Study of Desertification in Kordofan, Sudan, p. 61. Lund, Sweden: Lund Universitets Naturgeografiska Institution, mimeo.

Hellden, U. and Eklundh, L. 1988. *National Drought Impact Monitoring. A NOAA NDVI and Precipitation Data Study of Ethiopia*. Lund, Sweden: Lund University Press.

Hellinger, S., Hellinger, D., and O'Regan, F. 1988. *Aid for Just Development: Report on the Future of Foreign Assistance*. Boulder, Colo.: Lynne Rienner.

Helm, T. 1967. *Hurricanes: Weather at its Worst*. New York: Dodd/Mead.

Hervio, G. 1987. *Appraisal of Early Warning Systems in the Sahel* (Main Report). Paris: OECD/CILSS.

Hewitt, K. 1982. Settlement and Change in Basal Zone Ecotones: An Interpretation of the Geography of Earthquake Risk. In: B.G. Jones and M. Tomazevic (eds), *Social and Economic Aspects of Earthquakes*, pp. 15–42. Proceedings of the Third International Conference: The Social and Economic Aspects of Earthquakes and Planning to Mitigate their Impacts. Bled, Yugoslavia and Ithaca: Institute for Testing in Materials and Structures, Ljubljana and Cornell University.

—— (ed.) 1983a. *Interpretations Of Calamity*. Boston, Mass.: Allen & Unwin.

—— 1983b. The Idea of Calamity in a Technocratic Age. In: K. Hewitt (ed.), *Interpretations of Calamity*, pp. 3–32. Boston, Mass.: Allen & Unwin.

Holloway, R. 1989. *Doing Development – Governments, NGOs and the Rural Poor in Asia*. London: Earthscan.

Hopkins, R.F. 1987. The Evolution of Food Aid: Toward a Development-First Regime. In: J.P. Gittinger, J. Leslie, and C. Hoisington (eds), *Food Policy: Integrating Supply, Distribution, and Consumption*, pp. 246–59. Baltimore, Md.: Johns Hopkins University Press.

Horlick-Jones, T. 1990. *Acts of God? An Investigation Into Disasters*. London: London Emergency Planning Information Centre.

Horn, J. 1965. *Away with All Pests*. New York: Monthly Review.

Horowitz, M. 1989. Victims of Development. *Development Anthropology Network* 7,2: 1–8.

Horowitz, M. and Salem-Murdock, M. 1987. Political Economy of Desertification in White Nile Province, Sudan. In: P. Little and M. Horowitz (eds), *Lands at Risk in the Third World*, pp. 95–114. Boulder, Colo.: Westview.

—— 1990. Management of an African Floodplain: A Contribution to the Anthropology of Public Policy. In: M. Marchand and H. Udo de Haes (eds), *The People's Role in Wetland Management*, pp. 229–36. Leiden: Centre for Environmental Studies.

Hossain, H., Dodge, C.P., and Abed, F.H. (eds) 1992. *From Crisis to Development: Coping with Disasters in Bangladesh*. Dhaka: The University Press.

Hossain, M., Islam, A.T., and Samat Sana 1987. *Floods in Bangladesh: Recurrent Disaster and People's Survival*. Dhaka: Universities Research Centre.

Housner, G.W. 1989. *Coping with Natural Disasters, The International Decade for Natural Disaster Reduction*. London: SECAD.

Howard, J. and Lloyd, B. 1979. Sanitation and Diseases in Bangladesh Urban Slums and Refugee Camps. *Progress in Water Technology* 11: 191–200.

Hughes, C. and Hunter, J. 1970. Disease and 'Development' in Africa. *Social Science and Medicine* 3: 443–93.

Hussein, A.M. 1976. The Political Economy of Famine in Ethiopia. In: A. Hussein (ed.), *Rehab: Drought and Famine in Ethiopia*, pp. 9–43. London: International African Institute.

Ibrahim, F.N. 1991. The Exchange Rates of Livestock and Grain and Their Role in Enhancing Vulnerability to Famine in the Semi-Arid Zone of the Sudan. In: F.N. Ibrahim, H.G. Bohle, T. Cannon, and G. Hugo (eds), *Famine and Food Security in Africa and Asia*, pp. 185–9. Bayreuth: University of Bayreuth.

ICIHI (Independent Commission on International Humanitarian Issues) 1986. *The Encroaching Desert*. London: Zed Books.

—— 1988. *Winning the Human Race*. London: Zed Books.

Indra, D.M. and Buchignani, N. 1992. *Uthuli*. Residence as a Response to Environmentally-Forced Migration in Kazipur, Bangladesh. Paper presented to the Society for Applied Anthropology, 26 March, Memphis, Tennessee.

International Centre, Cities on Water 1989. *Impact of Sea Level Rise on Cities and Regions*. Venice: International Centre (S. Marco 875, 30124, Venice).

International Federation of Red Cross and Red Crescent Societies (FRCS) and Centre for Research in the Epidemiology of Disasters (CRED) 1993. *World Disasters Report*. Geneva: Federation of Red Cross and Red Crescent Societies (FRCS).

Isaza, P., de Quinteros, Z., Pineda, E., Parchment, C., Aguilar, E., and McQuestion, M. 1980. A Diarrheal Disease Control Programme Among Nicaraguan Refugee Children in Campo Luna, Honduras. *Bulletin of the Pan American Health Organization* 14: 337–42.

Islam, M.A. 1974. Tropical Cyclones: Coastal Bangladesh. In: G. White (ed.), *Natural Hazards*, pp. 19–25. New York: Oxford University Press.

Ives, J. and Messerli, B. 1989. *The Himalayan Dilemma: Reconciling Development and Conservation*. London: Routledge.

Ives, J. and Pitt, D.C. 1988. *Deforestation: Social Dynamics in Watersheds and Mountain Ecosystems*. London: Routledge.

Jackson, T. 1982. *Against the Grain: The Dilemma of Project Food Aid*. London: Oxfam.

Jacobs, D. 1987. *The Brutality of Nations*. New York: Paragon.

Jacobson, J. 1988. *Environmental Refugees: A Yardstick of Habitability*. Worldwatch Paper 86. Washington, DC: Worldwatch Institute.

James, L.D. and Pitman, K. 1992. The Flood Action Plan: Combining Approaches. *Natural Hazards Observer* 16,4: 6–7.

Jeffrey, S. 1980. Universalistic Statements About Human Social Behaviour. *Disasters* 4,1: 111–12.

—— 1982. The Creation of Vulnerability to Natural Disaster: Case Studies from the Dominican Republic. *Disasters* 6,1.

Jiggins, J. 1986. Women and Seasonality: Coping with Crisis and Calamity. *IDS Bulletin* 17,3: 9–18.

Jodha, N. 1991. Rural Common Property Resources: A Growing Crisis. Gatekeeper Series No. 24. London: International Institute for Environment and Development.

Johnston, B. and Schulte, J. 1992. Natural Power and Power Plays in Watsonville, California, and the U.S. Virgin Islands. Paper presented to the Society for Applied Anthropology, 26 March, Memphis, Tennessee.

Johnston, P. and Simmonds, M. 1991. Green Light for Precautionary Science. *New Scientist* 3 August: 4.

Juma, C. 1989. *The Gene Hunters*. London and Princeton, NJ: Zed Books and Princeton University Press.

Kalyalya, D., Mhlanga, K., Semboja, J., and Seidman, A. 1988. *Aid and Development in Southern Africa: A Participatory Learning Process*. Trenton, NJ: Africa World Press and Oxfam America.

Kameir, el W. and Karsany, I. 1985. *Corruption as the Fifth Factor of Production in the Sudan*. Report No. 2. Uppsala: Scandinavian Institute for African Studies.

Kane, P. 1988. *Famine in China: Demographic and Social Implications*. London: Macmillan.

Kapuscinski, R. 1983. *The Emperor: Downfall of an Autocrat*. London, Melbourne, and New York: Quartet Books.

Kebbede, G. 1992. *The Ethiopian Predicaments: State-Dictated Development, Ecological Crisis, Famine, and Mass Displacement*. Atlantic Hights, NJ: Humanities Press.

Kemp, P. 1991. For Generations to Come: The Environmental Catastrophe. In: P. Bennis and M. Moushabeck (eds), *Beyond the Storm: A Gulf Crisis Reader*, pp. 325–34. New York: Oliver Branch Press.

Kent, G. 1987. *Fish, Food, and Hunger*. Boulder, Colo.: Westview.

—— 1988. Nutrition Education as an Instrument of Empowerment. *Journal of Nutrition Education* 20,4: 193–5.

Kent, R.C. 1987a. *Anatomy of Disaster Relief: The International Network in Action*. London and New York: Pinter Publishers.

—— 1987b. Disaster Monitor. In: Raana Gauhar (ed.), *Third World Affairs 1987*, pp. 251–310. London: Third World Foundation for Social and Economic Studies.

Kerner, D.O. and Cook, K. 1991. Gender, Hunger and Crisis in Tanzania. In: R. Downs, D. Kerner, and S. Reyna (eds), *The Political Economy of African Famine*, pp. 257–72. Philadelphia: Gordon & Breach Science Publishers.

Khan, M. and Shahidullah, M. 1982. The Role of Water and Sanitation in the Incidence of Cholera in Refugee Camps. *Transactions of the Royal Society of Tropical Medicine and Hygiene* 76: 373–7.

Khan, M.I. 1991. The Impact of Local Elites on Disaster Preparedness Planning: The Location of Flood Shelters in Northern Bangladesh. *Disasters* 15,4: 340–54.

Khondker, H.H. 1992. Floods and Politics in Bangladesh. *Natural Hazards Observer* 16,4: 4–6.

Kibreab, G. 1985. *African Refugees*. Trenton, NJ: Africa World Press.

Kiljunen, K. (ed.) 1984. *Kampuchea: Decade of Genocide: Report of a Finnish Inquiry Commission*. London: Zed Press.

Kirby, A. (ed.) 1990a. *Nothing to Fear: Risks and Hazards in American Life*. Tucson, Ariz.: University of Arizona Press.

—— 1990b. On Social Representations of Risk. In: A. Kirby (ed.), *Nothing to Fear: Risks and Hazards in American Life*, pp. 1–16. Tucson, Ariz.: University of Arizona Press.

—— 1990c. Toward a New Risk Analysis. In: A. Kirby (ed.), *Nothing to Fear: Risks and Hazards in American Life*, pp. 281–98. Tucson, Ariz.: University of Arizona Press.

Kiriro, A. and Juma, C. (eds) 1989. *Gaining Ground: Institutional Innovations in Land-use Management in Kenya*. Nairobi: ACTS Press (African Centre for Technology Studies).

Kiser, L.J., Heston, J., Hickerson, S., Millsap, P., Nunn, W., and Pruitt, D. 1992. Anticipatory Stress in Children and Adolescents. Paper presented at the 51st Annual Meeting of the Society for Applied Anthropology, 26 March, Memphis, Tennessee.

257

Kjekshus, H. 1977. *Ecological Control and Economic Development in East African History*. Berkeley, Calif.: University of California Press.

Klee, G. (ed.) 1980. *World Systems of Traditional Resource Management*. New York: Halsted.

Kloos, H. 1982. Development, Drought, and Famine in the Awash Valley of Ethiopia. *African Studies Review* 25,4: 21–48.

Kreimer, A. and Echeverria, E. 1991. Case Study: Housing Reconstruction in Mexico City. In: A. Kreimer and M. Munasinghe (eds), *Managing Natural Disasters and the Environment*, pp. 53–61. Washington, DC: World Bank.

Kreimer, A. and Munasinghe, M. (eds) 1991. *Managing Natural Disasters and the Environment*. Washington, DC: World Bank.

Kreimer, A. and Zador, M. (eds) 1989. *Colloquium on Disasters, Sustainability and Development: A Look to the 1990s*. Environment Working Paper No. 23. Washington, DC: World Bank.

Kristof, N.D. 1991. In Bangladesh Storms, Poverty More than Weather is the Killer. *New York Times* 11 May: A1 and A5.

—— 1992. China's Floods of July: Misery Lingers. *New York Times* 27 January: A6.

Kruks, S. and Wisner, B. 1989. Ambiguous Transformations: Women, Politics and Production in Mozambique. In: S. Kruks, R. Rapp, and M. Young (eds), *Promissory Notes: Women in the Transition to Socialism*, pp. 148–71. New York: Monthly Review.

Kuester, I. and Forsyth, S. 1985. Rabaul Eruption Risk: Population Awareness and Preparedness Survey. *Disasters* 9,3: 179–82.

Kumar, G. 1987. *The Ethiopian Famine and Relief Measures: An Analysis and Evaluation*. Addis Ababa: UNICEF.

Laird, R. 1992. Private Troubles and Public Issues: The Politics of Disaster. Paper presented to the Society for Applied Anthropology, 26 March, Memphis, Tennessee.

Lamprey, M. 1976. *Survey of Desertification in Kardofan Province*. Nairobi: UNEP.

Langlands, B. (ed.) 1968. *The Medical Atlas of Uganda*. Kampala: Makerere University, Department of Geography.

Lappé, F.M., Collins J., and Kinley D. 1980. *Aid as Obstacle, Twenty Questions about Our Foreign Aid and the Hungry*. San Francisco, Calif.: Institute for Food and Development Policy.

Lawrence, P. (ed.) 1986. *World Recession and the Food Crisis in Africa*. London: John Currey.

Le Moigne, G., Barghouti, S., and Plusquellec, H. (eds) 1990. *Dam Safety and the Environment*. Technical Paper No. 115. Washington, DC: World Bank.

Leach, G. and Mearns, R. 1989. *Beyond the Woodfuel Crisis*. London: Earthscan.

Learmonth, A. 1988. *Disease Ecology*. Oxford: Basil Blackwell.

Leftwich, A. and Harvie, D. 1986. *The Political Economy of Famine*. Institute for Research in the Social Sciences. York: University of York.

Lemma, H. 1985. The Politics of Famine in Ethiopia. *Review of African Political Economy* 33: 44–58.

Lewis, J. 1981. Some Perspectives on Natural Disaster Vulnerability in Tonga. *Pacific Viewpoint* 22,2: 145–62.

—— 1984a. A Multi-Hazard History of Antigua. *Disasters* 8,3: 190–7.

—— 1984b. Disaster Mitigation Planning: Some Lessons from Island Countries. Occasional Paper, Centre for Development Studies. Bath: University of Bath.

—— 1987. Vulnerability and Development – and the Development of Vulnerability: A Case for Management. Development Studies Association, Annual Conference, University of Manchester, 16–18 September.

—— 1989. Sea-Level Rise: Tonga, Tuvalu (Kiribati). Commonwealth Expert Group on Climate Change and Sea Level Rise. London: Commonwealth Secretariat.

—— 1990. The Vulnerability of Small Island-States to Sea Level Rise: The Need for Holistic Strategies. *Disasters* 14,3.

Lewis, N.A. 1991. String of Crises Overwhelms Relief Agencies and Donors. *New York Times* 4 May: A5.

Linear, M. 1985. *Zapping the Third World: The Disaster of Development Aid.* London: Pluto Press.

Lipton, M. 1988. The Poor and the Poorest: Some Interim Findings. World Bank Discussion Paper 25. Washington, DC: World Bank.

Lipton, M. and Longhurst, R. 1989. *New Seeds and Poor People*, Baltimore, Md. and London: Johns Hopkins University Press and Unwin Hyman.

Lisk, F. (ed.) 1985. *Popular Participation in Planning for Basic Needs.* Aldershot: Gower.

Little, P. and Horowitz, M. (eds) 1987. *Lands at Risk in the Third World: Local-Level Perspectives.* Boulder, Colo.: Westview.

Liverman, Diana M. 1989. Vulnerability to Global Environmental Change. Paper presented to International Workshop on Understanding Global Environmental Change, Clark University, Center for Technology, Environment and Development, 11–13 October.

Longhurst, R. 1986. Household Food Strategies in Response to Seasonality and Famine. *IDS Bulletin* 17.

Lopez, M.E. 1987. The Politics of Lands at Risk in a Philippine Frontier. In P. Little and M. Horowitz (eds), *Lands At Risk in the Third World*, pp. 230–48. Boulder, Colo.: Westview.

Lovell, W.G. 1990. Maya Survival in Ixil Country, Guatemala. *Cultural Survival Quarterly* 14,4: 10–12.

Ludlum, D.M. 1963. *Early American Hurricanes 1492–1870.* Boston, Mass.: American Meteorological Society.

Lyngdoh, J.M. 1988. Disaster Management: A Case Study of Kosi Security System in North-East Bihar. *Journal of Rural Development* (Hyderabad) 7,5: 519–40.

McAlpin, M. 1983. *Subject to Famine: Food Crisis and Economic Change in Western India, 1860–1920.* Princeton, NJ: Princeton University Press.

McGlothlen, M.E., Goldsmith, P., and Fox, C. 1986. Undomesticated Animals and Plants. In: A. Hansen and D.E. McMillan (eds), *Food in Sub-Saharan Africa*, pp. 222–38. Boulder, Colo.: Lynne Rienner Publishers.

McGranahan, D., Pizarro, E., and Richard, C. 1985. *Measurement and Analysis of Socio-Economic Development.* Geneva: UNRISD.

McIntire, J. 1987. Would Better Information From an Early Warning System Improve African Food Security? In: D. Wilhite and W. Easterling (eds), *Planning for Drought*, pp. 283–93. Boulder, Colo.: Westview.

McKeown, T. 1988. *The Origins of Human Disease.* Oxford: Basil Blackwell.

MacMahon, B. and Pugh, T. 1970. *Epidemiology: Principles and Methods.* Boston, Mass.: Little, Brown.

McNeil, W.H. 1979. *Plagues and Peoples.* Harmondsworth: Penguin.

Mafeje, A. 1987. Food for Security and Peace in the SADCC Region. In: E. Hansen (ed.), *Africa: Perspectives on Peace and Development*, pp. 183–212. London: Zed Press.

Mahjoub, A. (ed.) 1990. *Adjustment or Delinking? The African Experience.* Tokyo and London: United Nations University Press and Zed Press.

Mahmud, A. 1988. Navies Hunt for Victims of Cyclone. *Guardian* (London) 2 December.

Mallory, W.H. 1926. *China: Land of Famine*. New York: American Geographical Publishing Society.

Maltby, E. 1985. *Peat Mining in Jamaica*. SIEP 2. Gland, Switzerland: IUCN.

—— 1986. *Waterlogged Wealth: Why Waste the World's Wet Places*. London: Earthscan.

Mamdani, M. 1985. Disaster Prevention: Defining the Problem. *Review of African Political Economy* 33: 92–6.

Marchand, M. and Udo de Haes, H.A. (eds) 1990. *The People's Role in Wetland Management*. Leiden: Centre for Environmental Studies.

Marglin, F. and Marglin, S. (eds) 1990. *Dominating Knowledge: Development, Culture, and Resistance*. Oxford: Clarendon Press.

Margolis, M. 1988. The Deadly Rains of Rio. *Newsweek* 7 March: 23.

Mariam, M.W. 1986. *Rural Vulnerability to Famine in Ethiopia 1958–1977*. London: Intermediate Technology Publications.

Marks, G. and Beatty, W. 1976. *Epidemics*. New York: Scribners.

Mascarenhas, A. 1971. Agricultural Vermin in Tanzania. In: S. Ominde (ed.), *Studies in the Geography and Development of East Africa*. Nairobi: Heinemann.

MASDAR (UK) Ltd 1987. *Resource Appraisal and Development Study of Selected Areas of North East Darfur*. Berkshire: MASDAR.

Maskrey, A. 1989. *Disaster Mitigation: A Community Based Approach*. Development Guidelines No. 3. Oxford: Oxfam.

Maskrey, A. and Romero, G. 1983. *Como Entender los Desastres Naturales*. Lima: PREDES.

Maslow, A. 1970. *Motivation and Personality*, 2nd edn. New York: Harper & Row.

Mason, R. 1992. The Awakening of Local Environmental Advocacy Following the Exxon Valdez Oil Spill in Kodiak, Alaska. Paper presented to the Society for Applied Anthropology, 26 March, Memphis, Tennessee.

Mass, W. 1970. *The Netherlands at War: 1940–1945*. New York: Aberlard-Schumann.

Matthiessen, C. 1992. The Day the Poison Stopped Working. *Mother Jones* March/April: 48–55.

Maybury, R.H. (ed.) 1986. *Violent Forces of Nature*. Mt Airy, Md.: Lomond Publications and UNESCO.

Maxwell, S. (ed.) 1991. *To Cure All Hunger. Food Policy and Food Security in Sudan*. London: Intermediate Technology Publications.

Mazumder, D. and Chakrabarty, A. 1973. Epidemic of Smallpox Among the Evacuees from Bangladesh in Salt Lake Area Near Calcutta. *Journal of Indian Medical Association* 60: 275–80.

Mbithi, P.M. and Wisner, B. 1973. Drought and Famine in Kenya: Magnitude and Attempted Solutions. *Journal of East African Research and Development* 3,2: 113–43.

Meillasoux, C. (ed.) 1973. *Qui se nourrit de la famine en Afrique?* Paris: Maspero.

—— 1974. Development or Exploitation: Is the Sahel Famine Good Business? *Review of African Political Economy* 1,1: 27–33.

Mellor, J.W. and Gavian, S. 1987. Famine, Causes, Prevention and Relief. *Science* 235: 539–45.

Merani, N.S. 1990. The International Decade for Natural Disaster Reduction. In: A. Kreimer and M. Munasinghe (eds), *Managing Natural Disasters and the Environment*, pp. 36–9. Washington, DC: World Bank.

Merchant, C. 1989. *The Death of Nature: Women, Ecology and the Scientific Revolution*. San Francisco, Calif.: Harper.

Messer, E. 1991. *Food Wars: Hunger as a Weapon of War in 1990*. Alan Shawn

Feinstein World Hunger Program Research Report. Providence, RI: Brown University.

Michaels, J. 1988. Rains Pour Torrent of Woe on Poor Rios Poor. *Christian Science Monitor* 6 March: 7.

Mileti, D.S., Drabek, T.E., and Haas, J.E. 1975. *Human Systems in Extreme Environments*. Monograph 21. Boulder, Colo.: University of Colorado Institute of Behavioral Science, Program on Environment and Behavior.

Miller, K.S. and Simile, C. 1992. They Could See Stars from Their Beds: The Plight of the Rural Poor in the Aftermath of Hurricane Hugo. Paper presented to the Society for Applied Anthropology, 26 March, Memphis, Tennessee.

Mills, C.W. 1959. *The Sociological Imagination*. New York: Oxford University Press.

Milne, A. 1986. *Floodshock: The Drowning of Planet Earth*. Gloucester: Alan Sutton.

Minear, L. 1991. *Operation Lifeline Sudan*. Trenton, NJ: Red Sea Press.

Mitchell, J.K. 1974. Community Response to Coastal Erosion: Individual and Collective Adjustments to Hazard on the Atlantic Shore. Research Paper 156. Chicago, Ill.: University of Chicago, Department of Geography.

—— 1985. Prospects for Improved Hurricane Protection on Oceanic Islands: Hawaii After Hurricane Iwa. *Disasters* 9,4: 286–94.

—— 1987. A Management-Oriented, Regional Classification of Developed Coastal Barriers. In: R.H. Platt (ed.), *Cities on the Beach*, pp. 31–42. Research Paper 224. Chicago, Ill.: University of Chicago, Department of Geography.

—— 1990. Human Dimensions of Environmental Hazards. In: A. Kirby (ed.), *Nothing to Fear*, pp. 131–75. Tucson, Ariz.: University of Arizona Press.

Momsen, J.H. and Townsend, J. (eds) 1987. *Geography of Gender in the Third World*. London: Hutchinson.

Monan, J. 1989. *Bangladesh: The Strength to Succeed. A Report for Oxfam*. Oxford: Oxfam.

Morgan, R. 1988. Drought-Relief Programmes in Botswana. In: D. Curtis, M. Hubbard, and A. Shepherd (eds), *Preventing Famine: Policies and Prospects for Africa*, pp. 112–20. London: Routledge.

Morris, J., West, G., Holck, S., Blake, P., Echeverria, P., and Karaulnik, M. 1982. Cholera Among Refugees in Rangsit, Thailand. *Journal of Infectious Diseases* 1: 131–4.

Morris, R. and Sheets, H., 1974. *Disaster in the Desert*. Special Report. Washington, DC: Carnegie Endowment for International Peace.

Mortimore, M. 1989. *Adapting to Drought: Farmers, Famines and Desertification in West Africa*. Cambridge: Cambridge University Press.

Muhema, B. 1972. The Impact of Flooding in Rufiji. *Journal of the Geographical Association of Tanzania* 7: 49–64.

Munasinghe, M., Menezes, B., and Preece, M. 1991. Case Study: Rio Flood Reconstruction and Prevention Project. In: A. Kreimer and M. Munasinghe (eds), *Managing Natural Disasters and the Environment*, pp. 28–31. Washington, DC: World Bank.

Murphy, L.M. and Moriarty, A.B. 1976. *Vulnerability, Coping and Growth from Infancy to Adolescence*. New Haven, Conn.: Yale University Press.

Murray, C. 1981. *Families Divided: The Impact of Migrant Labour in Lesotho*. Cambridge: Cambridge University Press.

Murray, M.J., Murray, A., Murray, N., and Murray, M.B. 1978. Diet and Cerebral Malaria: The Effect of Famine and Refeeding. *American Journal of Clinical Nutrition* 31: 57–61.

Nafziger, E. 1988. *Inequality in Africa*. Cambridge: Cambridge University Press.

National Oceanic and Atmospheric Administration (NOAA) 1980. *Significant Earthquakes 1900–1979, Map and Listing*. Washington, DC: NOAA.

Nelson, R. 1988. *Dryland Management: The Desertification Problem*, Working Paper No. 8. Washington, DC: World Bank.

Nerfin, M. 1990. Environment and Development: Listen to the South Citizen. *IFDA Dossier 77*, May June: 1 2. Nyon, Switzerland: International Foundation for Development Alternatives.

New Scientist. 1989. Editorial 28 October: 3.

Newbury, D. 1986. From 'Frontier' to 'Boundary': Some Historical Roots of Peasant Strategies of Survival in Zaire. In: G. Nzongola-Ntalaja (ed.), *The Crisis in Zaire*, pp. 87–112. Trenton, NJ: Africa World Press.

Newell, K. 1988. Selective Primary Health Care: The Counter Revolution. *Social Science and Medicine* 26,9: 903–6.

Newhall, C. 1993. Conversation with Ian Davis on the evacuation before the Pinatubo volcanic eruption.

Newman, L.F. (ed.) 1990. *Hunger in History: Food Shortage, Poverty and Deprivation*. Oxford: Basil Blackwell.

Newman, S. 1989. Earthweek: Diary of the Planet (Week Ending 13 October). *San Francisco: Chronicle* Features.

Nicaragua Ecumenical Group 1988. Statement on Hurricane Joan. *Lucha/Struggle: A Journal of Christian Reflection on Struggles for Liberation* 12,6: 38–9.

Nichols, N. 1988. Food Information Systems in Sub-Saharan Africa: Effective Tools or Illusions of Preparedness? M.Sc. Thesis, School of Development Studies, University of East Anglia.

Noble, J.H. 1981. Social Inequity in the Prevalence of Disability. *Assignment Children* 53–4: 23–32.

NRC (US National Research Council) 1991. *A Safer Future: Reducing the Impacts of Natural Disasters*. Washington, DC: National Research Council.

Nuguid, A.P. 1990. Environmental Abuse Getting Worse. *Business Star* (Bangkok) 2 March.

O'Brien, C. and O'Brien, M. 1972. *The Story of Ireland*. New York: Viking.

O'Brien, J. 1980. Agricultural Labor and Development in Sudan. Ph.D. Thesis, University of Connecticutt.

—— 1983. Formation of the Agricultural Labor Force. *Review of African Political Economy* 26: 15–34.

O'Brien, J. and Gruenbaum, E. 1991. A Social History of Food, Famine and Gender in Twentieth-Century Sudan. In: R. Downs, D. Kerner, and S. Reyna (eds), *The Political Economy of African Famine*, pp. 177–203. Philadelphia: Gordon & Breach Science Publishers.

Odegi-Awuoundo, C. 1990. *Life in the Balance: Ecological Sociology of Turkana Nomads*. Nairobi: ACTS Press.

Odhiambo, T., Anyang'Nyong'o, P., Hansen, E., Lardner, G., and Wai, D. (eds) 1988. *Hope Born out of Despair: Managing the African Crisis*. Nairobi: Heinemann Kenya.

Office of Foreign Disaster Assistance (OFDA) 1992. *OFDA Annual Report* 1991. Washington, DC: Office of Foreign Disaster Assistance, Agency for International Development.

O'Keefe, P., Westgate, K., and Wisner, B. 1977. Taking the Naturalness Out of Natural Disasters. *Nature* 260, 15 April: 566–7.

O'Keefe, P. and Wisner, B. 1975. African Drought: The State of the Game. In: P.

Richards (ed.), *African Environment: Problems and Perspectives*, pp. 31–9. London: International African Institute.

—— (eds) 1977. *Landuse and Development*. African Environment Special Report no. 5. London: International African Institute.

Oliver-Smith, A. 1986a. Disaster Context and Causation: An Overview of Changing Perspectives in Disaster Research. In: A. Oliver-Smith (ed.), *Natural Disasters and Cultural Responses*, pp. 1–34. Studies in Third World Societies No. 36. Williamsburg, Va.: College of William and Mary.

—— 1986b. *The Martyred City: Death and Rebirth in the Andes*. Albuquerque, NM: University of New Mexico Press.

—— 1988. Planning Goals and Urban Realities: Post-Disaster Reconstruction in a Third World City. *City and Society* 2,2: 105–26.

—— 1990. Post-Disaster Housing Reconstruction and Social Inequality – A Challenge to Policy and Practice. *Disasters* 14,1: 7–19.

—— 1992. Remarks as Discussant Following Panel, *The Politics of Disaster*, Annual Meeting of the Society for Applied Anthropology, 26 March, Memphis, Tennessee.

—— 1994. The Five Hundred Year Earthquake: Natural and Social Hazards in the Third World (Peru). In: A. Varley (ed.), *Disasters, Development and the Environment*. London: Belhaven Press.

O'Neill, B. 1990. Cities Against the Seas. *New Scientist* 125,1702: 3.

Onimode, B. (ed.) 1989. *The IMF, the World Bank, and the African Debt*, 2 vols. London: Zed Press and the Institute for African Alternatives.

O'Riordon, T. 1986. Coping with Environmental Hazards. In: R. Kates and I. Burton (eds), *Geography, Resources, and Environment*, vol. 2, pp. 272–309. Chicago, Ill.: University of Chicago Press.

Oxfam 1988. Debt Crisis Case Study: Jamaica. People in Crisis Campaign Leaflet. Oxford: Oxfam.

Pacific Islands Development Program n.d. *Agricultural Development and Disaster Preparedness*. Honolulu: East–West Center.

Packard, R. 1989. *White Plague, Black Labor: Tuberculosis and the Political Economy of Health in South Africa*. Berkeley, Calif.: University of California Press.

Packard, R. and Epstein, R. 1987. Ecology and Immunology: The Social Science Context of AIDS in Africa. *Science for the People* 19,1: 10–17.

Packard, R., Wisner, B., and Bossart, T. (eds) 1989. Political Economy of Health and Disease in Africa and Latin America. Special issue of *Social Science and Medicine* 28,5: 405–530.

Palm, R.I. 1990. *Natural Hazards: An Integrative Framework for Research and Planning*, Baltimore, Md.: Johns Hopkins University Press.

Pan American Health Organization (PAHO) 1982. *Epidemiologic Surveillance after Natural Disaster*. Washington, DC: Pan American Health Organization.

Pankhurst, R. 1974. *The History of Famine and Epidemics in Ethiopia Prior to the Twentieth Century*. Addis Ababa: Relief and Rehabilitation Commission.

PANOS 1989. *AIDS and the Third World*. London: Panos Institute and Norwegian Red Cross.

Parker, D.J. 1992. The Flood Action Plan: Social Impacts in Bangladesh. *Natural Hazards Observer* 16,4: 3–4.

Parker, R.S. 1989. Proyecto Nueva Vida Armero. In: M.B. Anderson and P.J. Woodrow (eds), *Rising from the Ashes, Development Strategies in Times of Disaster*, pp. 159–83. Boulder, Colo.: Westview.

Parr, A.R. 1987. Disaster and Disabled Persons: An Examination of the Safety Needs of a Neglected Minority. *Disasters* 11: 148–53.

263

Parry, M.L. and Carter, T.R. 1987. Climate Impact Assessment: A Review of Some Approaches. In: D. Wilhite and W. Easterling (eds), *Planning for Drought: Toward a Reduction of Societal Vulnerability*, pp. 165–87. Boulder, Colo.: Westview.

Paul, B.K. 1984. Perception of and Agricultural Adjustments to Floods in Jamuna Floodplain, Bangladesh. *Human Ecology* 12,1: 3–19.

Pavlovsky, Y. (ed.) n.d. *Human Diseases with Natural Foci*. Moscow: Foreign Languages Publishing House.

Pearce, D. 1987. *Natural Resources Management in West Sudan*. Khartoum: World Bank.

Pearce, D., Barbier, E., and Markandya, A. 1990. *Sustainable Development: Economics and Environment in the Third World*. London: Earthscan.

Pearce, D., Markandya, A., and Barbier, E. 1989. *Blueprint for a Green Economy*. London: Earthscan.

Pearce, D. and Turner, R.K. 1990. *Economics of Natural Resources and the Environment*. Baltimore, Md.: Johns Hopkins University Press.

Pearce, F. 1989. *Turning up the Heat: Our Perilous Future in the Global Greenhouse*. London: Paladin.

—— 1991. The Rivers That Won't be Tamed. *New Scientist* 13, April: 38–41.

Pelanda, C. 1981. Disaster and Socio-Systemic Vulnerability. Third International Conference on the Social and Economic Aspects of Earthquakes and Planning to Mitigate their Impacts. Bled, Yugoslavia.

Penning-Rowsell, E.C., Parker, D.J., and Harding, D.M. 1986. *Floods and Drainage: British Policies for Hazard Reduction, Agricultural Improvement and Wetland Conservation*. London: Allen & Unwin.

Perrow, C. 1984. *Normal Accidents: Living with High Risk Technologies*. New York: Basic Books.

Perry, R. and Mushkatel, A., 1986. *Minority Citizens in Disasters*. Athens, Ga.: University of Georgia Press.

Petak, W.J. and Atkisson, A.A. 1982. *Natural Hazard Risk Assessment and Public Policy*. New York: Springer Verlag.

Peterson, M. (ed.) 1977. *The Portable Thomas Jefferson*. London: Penguin.

Philippine Institute of Volcanology and Seismology n.d.(a). *Operation Taal*. Quezon City: Phivolus Press.

—— n.d.(b). *Volcanoes and Philippine Volcanology*. Quezon City: Philippine Institute of Volcanology and Seismology.

Pinstrup-Anderson, P. 1985. The Nutritional Effects of Export Crop Production: Current Evidence and Policy Implications. In: M. Biswas and P. Pinstrup-Anderson (eds), *Nutrition and Development*, pp. 43–59. Oxford: Oxford University Press.

—— (ed.) 1988. *Food Subsidies in Developing Countries: Costs, Benefits, and Policy Options*. Baltimore, Md.: Johns Hopkins University Press and IFPRI.

Plant, R. 1978. *Guatemala: A Permanent Disaster*. London: Latin America Bureau.

Platteau, J. 1988. *The Food Crisis in Africa: A Comparative Structural Analysis*. WIDER Working Paper No. 44. Helsinki: World Institute for Development Economic Research (WIDER).

Plessis-Fraissard, M. 1989. Mitigation Efforts at the Municipal Level: The La Paz Municipal Development Project. In: A. Kreimer and M. Zador (eds), *Colloquium on Disasters, Sustainability and Development: A Look to the 1990s*, pp. 132–5. Washington, DC: World Bank.

Poore, D. 1989. *No Timber Without Trees: Sustainability in the Tropical Forest*. London: Earthscan.

Popkin, R. 1990. The History and Politics of Disaster Management in the United States. In: A. Kirby (ed.), *Nothing to Fear*, pp. 101–30. Tucson, Ariz.: University of Arizona Press.

Porter, P. 1965. Environmental Potentials and Economic Opportunities: A Background for Cultural Adaptation. *American Anthropologist* 67: 409–20.

—— 1979. *Food and Development in the Semi-Arid Zone of East Africa*. Foreign and Comparative Studies/African Studies no. 32. Syracuse, NY: Maxwell School of Citizenship and Public Service, Syracuse University

Pradervand, P. 1989. *Listening to Africa: Developing Africa from the Grassroots*. New York: Praeger.

Prah, K.K. (ed.) 1988. *Food Security in Southern Africa*. Southern African Studies Series No. 4. Roma, Lesotho: Institute of Southern African Studies, National University of Lesotho.

Pratt, B. and Boyden, J. 1990. Personal communication with Ian Davis.

PRC (People's Republic of China), State Statistical Bureau 1986. *Statistical Yearbook of China 1986*. Hong Kong: Oxford University Press.

Press, F. 1990. Point of View. *UNESCO Sources* 11: 3.

Prindle, P.H. 1979. Peasant Society and Famines: A Nepalese Example. *Ethnology* 1.

Prothero, M. 1965. *Migrants and Malaria*. London: Longman.

Pryer, J. and Crook, N. 1988. *Cities of Hunger: Urban Malnutrition in Developing Countries*. Oxford: Oxfam.

Pryor, L. 1982. *Ecological Mismanagement in Natural Disasters*. Commission on Ecology Papers no. 2. Gland, Switzerland: IUCN Commission on Ecology/League of Red Cross Societies.

Pyle, A. and Gabbar, O.A. 1990. Household Vulnerability to Famine: Survival and Recovery Strategies Among Zasghawa and Berti Migrants in Northern Darfur, Sudan, 1982–1989. The Project on African Agriculture, Working Paper No. 2. New York: Social Science Research Council.

Quarantelli, E.L. (ed.) 1978. *Disasters: Theory and Research*. Sage Studies in International Sociology 13. Beverly Hills, Calif.: Sage Publications.

—— 1984. Perceptions and Reactions to Emergency Warnings of Sudden Hazards. *Ekistics* 51,309: 511–15.

—— 1990. Disaster Prevention and Mitigation in Lada: Problems and Options in Planning and Implementing in a Composite Country. Unpublished paper presented at Colloquium on the Environment and Natural Disaster Management. Washington, DC: World Bank.

Quarantelli, E.L. and Dynes, R.R. 1972. Images of Disaster Behavior: Myths and Consequences. Preliminary Paper 5. Columbus, OH: Disaster Research Center, Ohio State University.

—— 1977. Response to Social Crisis and Disaster. *Annual Review of Sociology* 3: 23–49.

Rahmato, D. 1988. *Peasant Survival Strategies*. Geneva: International Institute for Relief and Development/Food for The Hungry International.

Raikes, P. 1988. *Modernizing Hunger*. London: James Currey.

Ramos-Jimenez, P., Chiong-Javier, M.E., and Sevilla, J.C. 1986. *Philippine Urban Situation Analysis*. Manila: UNICEF.

Rangasami, A. 1985. 'Failure of Exchange Entitlements' Theory of Famine: A Response. *Economic and Political Weekly* 20,41, and 42, 12 and 19 October 1989.

—— 1986. Famine: The Anthropological Account; An Evaluation of the Work of Raymond Firth. *Economic and Political Weekly* 21,36: 1591–601.

Rao, N.V.K. 1974. Impact of Drought on the Social System of a Telengana Village. *Eastern Anthropologist* 27,4: 299–315.

Raphael, B. 1986. *When Disaster Strikes*. London: Hutchinson.

Rashid, 1977. *Geography of Bangladesh*. Dhaka: University Press.

Rashtriya Barh Ayog (National Commission on Floods) 1980. *Report*. 2 vols. New Delhi: Ministry of Energy and Irrigation.

Rau, B. 1991. *From Feast to Famine: Official Cures and Grassroots Remedies to Africa's Food Crisis*. London: Zed Press.

Ravallion, M. 1985. The Performance of Rice Markets in Bangladesh During the 1974 Famine. *Economic Journal* 95: 15–29.

—— 1987. *Markets and Famines*. Oxford: Clarendon Press.

Reacher, M., Campbell, C., Freeman, J., Doberstyn, E., and Brandling-Bennett, A. 1980. Drug Therapy for Plasmodium Falciparum Malaria Resistant to Pyrimethamine-Sulfadoxine (Fansidar): A Study of Alternative Regimens in Eastern Thailand. *Lancet* 3: 1066–9.

Read, B. 1970. *Healthy Cities: A Study of Urban Hygiene*. Glasgow: Blackie.

Reddy, A.V.S. 1991. Unpublished notes of a presentation on the Andhra Pradesh cyclone of 1990 to a Workshop on Disaster Management, June 1991 by the Director of the Centre for Disaster Management, Hyderabad. Oxford: Disaster Management Centre.

Regan, C. 1983. Underdevelopment and Hazards in Historical Perspective: An Irish Case Study. In: K. Hewitt (ed.), *Interpretations of Calamity*, pp. 98–120. Boston, Mass. and London: Allen & Unwin.

Reisner, M. 1986. *Cadillac Desert: The American West and its Disappearing Water*. New York: Penguin.

Richards, P. 1975. 'Alternative' Strategies for the African Environment: 'Folk Ecology' as a Basis for Community Oriented Agricultural Development. In: P. Richards (ed.), *African Environment: Problems and Perspectives*, pp. 102–17. London: International African Institute.

—— 1983. Ecological Change and the Politics of African Land Use. *African Studies Review* 26: 1–72.

—— 1985. *Indigenous Agricultural Revolution*. London: Hutchinson Education.

—— 1986. What's Wrong with Farming Systems Research? Paper presented at the conference of the Development Studies Association, University of East Anglia, Norwich.

Richards, P.J. and Thomson, A.M. 1984. *Basic Needs and the Urban Poor*. London: Croom Helm.

Rivers, J. 1982. Women and Children Last: An Essay on Sex Discrimination. *Disasters* 6,4: 256–67.

—— 1987. Famine Forecasting: Prices and Peasant Behaviour in Northern Ethiopia. *Disasters* 11.

Rivers, J., Holt, J., Seaman, J., and Bowden, M. 1974. Lessons for Epidemiology from the Ethiopian Famine. *Annales de la Société Belge de médecine tropicale* 56: 345–67.

ROAPE (*Review of African Political Economy*) 1985. *War and Famine*, theme issue, 33.

—— 1990. *What Price Economic Reform?*, theme issue, 47.

Rob, M.A. 1990. Flood Hazard in Bangladesh: Nature, Causes and Control. *Asian Profile* 18,4: 365–78.

Robinson, S., Franco, Y., Castrejon, R., and Bernard, H. 1986. It Shook Again: The Mexico City Earthquake of 1985. In: A. Oliver-Smith (ed.), *Natural Disasters and Cultural Responses*, pp. 81–122. Studies in Third World Societies No. 36. Williamsburg, Va.: College of William and Mary.

266

Robson, J.R.K. (ed.) 1981. *Famine: Its Causes, Effects and Management*. New York: Gordon & Breach

Rogers, B. 1980. *The Domestication of Women*. London: Tavistock.

Rogers, P., Lydon, P., and Seckler, D. 1989. *Eastern Waters Study: Strategies to Manage Flood and Drought in the Ganges-Brahmaputra Basin*. Report prepared by Irrigation Support Project for Asia and the Near East. Washington, DC: USAID.

Rogge, J.R. and Elahi, K.M. 1989. *The Riverbank Erosion Impact Study Bangladesh*. Final Report to the International Development Research Centre. Ottawa: IDRC.

Ross, L. 1984. Flood Control Policy in China: The Policy Consequences of Natural Disasters. *Journal of Public Policy* 3,2: 209–32.

Roundy, R.W. 1983. Altitudinal Mobility and Disease Hazards for Ethiopian Populations. *Economic Geography* 52: 103–15.

Roy, A. 1989. Greenhouse Hots up in a Hurry. *Sunday Times* 30 July.

Royal Academy of Engineering 1993. *Opportunities for British Involvement in the International Decade for Natural Disaster Reduction (IDNDR)*. Proceedings of Workshop, 27 March 1992. London: Royal Academy of Engineering.

Royal Society 1992. *Risk: Analysis, Perception and Management*. Report of a Royal Society Study Group. London: Royal Society.

Ruffié, J. 1987. *The Population Alternative: A New Look at Competition and the Species*. London: Penguin.

Russell, J. 1968. That Earlier Plague. *Demography* 5.

Saarinen, T.F. and Sell, J.L. 1985. *Warning and Response to the Mount St. Helen's Eruption*. Albany, NY: State University of New York Press.

Sabatier, R. 1988. *Blaming Others: Prejudice, Race and Worldwide AIDS*. London: Panos Institute.

Sagov, M. 1981. The Interface Between Earthquake Planning and Development Planning: A Case Study and Critique of the Reconstruction of Huaraz and the Callejou de Huaylas, Peru, Following the 31 May 1970 Earthquake. In: I. Davis (ed.), *Disaster and the Small Dwelling*. Oxford: Pergamon Press.

Sanchez, E., Cronick, K., and Wiesenveld, E. 1988. Psychological Variables in Participation: A Case Study. In: D. Canter, M. Krampen, and D. Stea (eds), *Ethnoscape*. Aldershot: Gower.

Sandberg, A. 1973. Ujamaa and Control of the Environment. Paper presented at the Annual Social Science Conference of East African Universities, Dar es Salaam, Tanzania.

Sapir, D. and Lechat, M.L. 1986. Reducing the Impact of Natural Disasters: Why Aren't We Better Prepared? *Health Policy and Planning* 1,2: 118–26.

Sattaur, O. 1991. Counting the Cost of Catastrophe. *New Scientist* 29 June: 21–3.

Schneider, S.H. 1989. *Global Warming: Are We Entering the Greenhouse Century?* San Francisco, Calif.: Sierra Club Books.

Schoepf, B.G. 1992. Gender Relations and Development: Political Economy and Culture. In: A. Seidman and F. Anang (eds), *21st Century Africa: Towards a New Vision of Self-Sustainable Development*, pp. 203–41. Atlanta, Ga. and Trenton, NJ: African Studies Association and Africa World Press.

Schoepf, B.G. and Schoepf, C. 1990. Gender, Land, and Hunger in Eastern Zaire. In: R. Huss-Ashmore and S. Katz (eds), *African Food Systems in Crisis: Contending with Change*, pp. 75–106. Philadelphia: Gordon & Breach Science Publishers.

Schroeder, R. 1987. *Gender Vulnerability to Drought: A Case Study of the Hausa Social Environment*. Natural Hazards Working Paper No. 58. Boulder, Colo.: University of Colorado, Institute of Behavioral Science.

Scott, J.C. 1976. *The Moral Economy of the Peasant: Rebellion and Subsistence in Southeast Asia*. New Haven, Conn. and London: Yale University Press.

—— 1985. *Weapons of the Weak: Everyday Forms of Peasant Resistance*. New Haven, Conn.: Yale University Press.

—— 1990. *Domination and the Arts of Resistance: Hidden Transcripts*. New Haven, Conn.: Yale University Press.

Scott, M. 1987. The Role of Non-governmental Organizations in Famine Relief and Prevention. In: M. Glantz (ed.), *Drought and Hunger in Africa*, pp. 349–66. Cambridge: Cambridge University Press.

Scott, M. and Mpanya, M. 1991. *We are the World: An Evaluation of Pop Aid for Africa*. Petaluma, Calif.: World College West and USA for Africa.

Scrimshaw, N., Gordon, J., and Taylor, C. 1968. *The Interaction of Nutrition and Infection*. WHO Technical Monograph 27. Geneva: World Health Organization.

Scudder, B. 1990. Energy Galore. *Geographical Magazine* September: 40–4.

Scudder, T. 1980. River-Basin Development and Local Initiative in African Savanna Environments. In: D. Harris (ed.), *Human Ecology in Savanna Environments*, pp. 383–406. London: Academic Press.

—— 1989. Conservation vs. Development: River Basin Projects in Africa. *Environment* 31,2: 4–9, 27–32.

Seager, J. 1992. Operation Desert Disaster: Environmental Costs of the War. In: C. Peters (ed.), *Collateral Damage: The 'New World Order' at Home and Abroad*, pp. 197–216. Boston, Mass.: South End.

Seale, J. 1991. Letter to the *Sunday Times*, 25 March.

Seaman, J. and Holt, J. 1980. Markets and Famines in the Third World. *Disasters* 4,3: 283–97.

Seaman, J., Leivesley, S., and Hogg, C. 1984. *Epidemiology of Natural Disasters*. Basle: Karger.

Sen, A. 1981. *Famines and Poverty*. London: Oxford University Press.

—— 1983. Development: Which Way Now? *Economic Journal* 93: 745–62.

—— 1985. *Food, Economics and Entitlements*. WIDER WP-1. 28 August.

—— 1988. Family and Food: Sex Bias in Poverty. In: T.N. Srinivasan and P.K. Bardhan (eds), *Rural Poverty in South Asia*, pp. 453–72. New York: Columbia University Press.

—— 1990. Gender and Cooperative Conflict. In: I. Tinker (ed.), *Persistent Inequalities: Women and World Development*, pp. 123–49. Oxford: Oxford University Press.

Sen, G. and Grown, C. 1987. *Development, Crisis, and Alternative Visions*. New York and London: Monthly Review and Earthscan.

Seth, S.L., Das, D.C., and Gupta, G.P. 1981. Floods in Arid and Semi-Arid Areas – Rajasthan. New Delhi: Ministry of Agriculture, mimeo.

Shah, B.V. 1983. Is the Environment Becoming More Hazardous: A Global Survey 1947–1980. *Disasters* 7,3: 202–9.

Shaker, M.T. 1987. An Analysis of Squatter Settlements in Dhaka, Bangladesh. Ph.D. Thesis, University of Liverpool.

Shakow, D. and O'Keefe, P. 1981. Yes, We Have No Bananas: The Economic Effects of Minor Hazards in the Windward Islands. *Ambio* 10,6: 344.

Shakur, T. 1987. An Analysis of Squatter Settlements in Dhaka, Bangladesh. Ph.D. Thesis, University of Liverpool.

Shanin, T. (ed.) 1971. *Peasants and Peasant Society*. Harmondsworth: Penguin.

Sharma, V.P. and Mehrotra, K.N. 1986. Malaria Resurgence in India: A Critical Study. *Social Science and Medicine* 22,2: 835–45.

Sharp, J.S. and Spiegel, A.D. 1984. Vulnerability to Impoverishment in South

African Rural Areas: The Erosion of Kinship and Neighborhood as Social Resources. Carnegie Conference Paper 52. Cape Town University.

Shepherd, A.W. 1984. Nomads Farmers and Merchants: Old Strategies in a Changing Sudan. In: E. Scott (ed.), *Life Before the Drought*, pp. 77–100. London: Allen & Unwin.

—— 1988. Case Study of Famine: Sudan. In: D. Curtis, M. Hubbard, and A. Shepherd (eds), *Preventing Famine: Policies and Prospects for Africa*, pp. 28–72. London: Routledge.

Shindo, E. 1985. Hunger and Weapons: The Entropy of Militarisation. *Review of African Political Economy* 33, August.

Shiva, V. 1989. *Staying Alive: Women, Ecology and Development*. London: Zed Press.

—— 1991. *The Violence of the Green Revolution: Third World Agriculture, Ecology and Politics*. London and Penang: Zed Press and Third World Network.

Sidel, R. and Sidel, V.W. 1982. *The Health of China: Current Conflicts in Medical and Human Services for One Billion People*. Boston, Mass.: Beacon Press.

Siegel, S.R. and Witham, P. 1991. Case Study Colombia. In: A. Kreimer and M. Munasinghe (eds), *Managing Natural Disaster and the Environment*, pp. 170–1. Washington, DC: World Bank.

Sigurdson, H. 1988. Gas Burst from Cameroon Crater Lakes: A New Natural Hazard. *Disasters* 12,2: 131–46.

Sigurdson, H. and Carey, S. 1986. Volcanic Disasters in Latin America and the 13th November 1985 Eruption of Nevado del Ruiz Volcano in Colombia. *Disasters* 10,3: 205–16.

Sikander, A.S. 1983. Floods and Families in Pakistan – A Survey. *Disasters* 7,2: 101–6.

Silgado, S. and Giesecke, A. 1983. *Terremotos en el Peru*. Lima: Ediciones Richay.

Simmonds, S., Vaughan, P., and Gunn, S. 1983. *Refugee Community Health Care*. Oxford: Oxford University Press.

Simpson, R.H. and Reidl, H. 1981. *The Hurricane and Its Impact*. Oxford: Basil Blackwell.

Sin, M. and Davies, J. 1994. The Khartoum Region's Vulnerability to Hazards and Disasters. In: A. Varley (ed.), *Disasters, Development and the Environment*. London: Belhaven Press.

SINAL (Boletim Imformativo do Sinal/Sistema de Informacoes a Nivel Local) 1992. O Rio e as enchentes. (Rio and floods). *SINAL* Nov.–Dec. 1991–Jan. 1992: 3.

Singer, H., Wood, J., and Jennings, T. 1987. *Food Aid: The Challenge and the Opportunity*. Oxford: Clarendon Press.

Singh, S.K. 1975. *The Indian Famine, 1967*. New Delhi: People's Publishing House.

SIPRI (Stockholm International Peace Research Institute) 1976. *Ecological Consequences of the Second Indochina War*. Stockholm: Almqvist & Wiksell International.

—— 1980. *Warfare in a Fragile World: Military Impact on the Human Environment*. London: Taylor & Francis.

Sjöberg, L. (ed.) 1987. *Risk and Society: Studies in Risk Generation and Reactions to Risk*. London: Allen & Unwin.

Slim, H. and Mitchell, J. 1990. Towards Community Managed Relief: A Case Study from Southern Sudan. *Disasters* 14,3: 265–8.

Smith, K. 1992. *Environmental Hazards: Assessing Risk and Reducing Disaster*. London: Routledge.

Smith, S. 1990. *Front Line Africa*. Oxford: Oxfam.

Sorokin, P.A. 1975. *Hunger as a Factor in Human Affairs*. Gainsville, Fla.: University of Florida Press.

Southern, R.L., The Global Socio-Economic Impact of Tropical Cyclones. *Australian Meteorological Magazine* 27: 175–95.

Spitz, P. 1976. *Famine-Risk and Famine Prevention in the Modern World: Studies in Food Systems Under Conditions of Recurrent Scarcity*. Geneva: United Nations Research Institute for Social Development (UNRISD).

Stark, K.P. and Walker, G.R. 1979. Engineering for Natural Hazards with Particular Reference to Tropical Cyclones. In: R.L. Heathcote and B.G. Thom (eds), *Natural Hazards in Australia*, pp. 189–203. Canberra: Australian Academy of Sciences.

Starosolszky, Oe. and Melder, O.M. (eds) 1989. *Hydrology of Disasters*. London: James & James Science Press.

Stevens, J.D. 1992. Fear and Loathing Along the New Madrid Fault System: The December 3, 1990 Earthquake Prediction Episode as a Learning Laboratory for the Dynamics of Information. Paper presented at the 51st Annual Meeting of the Society for Applied Anthropology, 26 March, Memphis, Tennessee.

Stewart, F. 1987. Should Conditionality Change? In: K. Havnevik (ed.), *The IMF and the World Bank in Africa: Conditionality, Impact and Alternatives*, pp. 29–46. Seminar Proceedings No. 18. Uppsala: Scandinavian Institute of African Studies.

Stock, R. 1976. *Cholera in Africa*. London: International African Institute.

Stoll, D. 1990. 'The Land No Longer Gives': Land Reform in Nebaj, Guatemala. *Cultural Survival Quarterly* 14,4: 4–9.

Susman, P., O'Keefe, P., and Wisner, B. 1983. Global Disasters, a Radical Interpretation. In: K. Hewitt (ed.), *Interpretations of Calamity*, pp. 274–6. London: Allen & Unwin.

Swift, J. 1989. Why are Rural People Vulnerable to Famine? *IDS Bulletin* 20,2: 8–15.

Tayag, J.C. n.d. (a) (c. 1985). How the People Escaped and Coped with Mayon Volcano's Fury – A Case Study of Institutional and Human Response to the 1984 Mayon Volcano Eruption. Quezon City: Philippine Institute of Volcanology and Seismology.

—— (ed.) n.d. (b.) (c. 1992). *Pinatubo Volcano Wakes from Four Century Slumber*. Philippine Institute of Volcanology and Seismology. Quezon City: Phivoles Press.

Temcharoen, P., Viboolyavatana, J., Tongkoom, B., Sumethanurugkul, B., Keittivuti, B., and Wanaratana, I. 1979. A Survey of Intestinal Parasitic Infections in Laotian Refugees at Ubon Province, With Special Reference to Schistosomiasis. *Southeast Asian Journal of Tropical Medicine and Public Health* 10: 552–4.

Thébaud, B. 1988. *Elevage et développement au Niger*. Geneva: ILO.

Thompson, M. 1989. The Role of Non-Government Agencies in Disaster Mitigation. Unpublished paper presented at Conference on the Role of NGOs in Disaster Mitigation. Disaster Management Centre, Oxford Polytechnic.

Thompson, M. and Warburton, M. 1988. Uncertainty on a Himalayan Scale. In: J. Ives and D. Pitt (eds), *Deforestation: Social Dynamics in Watersheds and Mountain Ecosystems*, pp. 1–53. London: Routledge.

Thompson, P. and Penning-Rowsell, E. 1994. Socio-Economic Impacts of Floods and Flood Protection: A Bangladesh Case Study. In: A. Varley (ed.), *Disasters, Development and the Environment*. London: Belhaven Press.

Tickell, C. 1990. *Climate Change and Development*. The 1989 Gilbert Murray Memorial Lecture. Oxford: Oxfam.

Tierney, K.J. 1992. Politics, Economics, and Hazards. Paper presented to the Society for Applied Anthropology, 26 March, Memphis, Tennessee.

Timberlake, L. 1985. *Africa in Crisis*. London: Earthscan.

Time-Life (eds) 1983. *Planet Earth, Volcano*. Amsterdam: Time-Life Books.

Timmerman, P. 1981. *Vulnerability, Resilience and the Collapse of Society*. Environmental Monograph No. 1, Institute for Environmental Studies. Toronto: University of Toronto.

Tinker, T. (ed.) 1990. *Persistent Inequalities: Women and World Development*. Oxford: Oxford University Press.

Tobriner, S. 1988. The Mexico Earthquake of September 19, 1985: Past Decisions, Present Danger: An Historical Perspective on Ecology and Earthquakes in Mexico City. *Earthquake Spectra* 4,3: 469–79.

Togolese Federation of Women in the Legal Profession 1988. Women's Participation in Development: The Case of Togo. In: K. Young (ed.), *Women and Economic Development*, pp. 171–208. Oxford and Paris: Berg and UNESCO.

Tomblin, J. 1981. Earthquakes, Volcanoes and Hurricanes: A Review of Natural Hazards and Vulnerability in the West Indies. *Ambio* 10,6: 340–5.

—— 1985. Armero: The Day Before. *UNDRO NEWS* Nov.–Dec.: 4–6.

—— 1987. Management of Volcanic Emergencies. *UNDRO NEWS* July–August: 17.

Torry, W.I. 1986. Economic Development, Drought, and Famine. Some Limitations of Dependency Explanations. *Geojournal* 12,1: 5–18.

Trainer, T. 1989. *Developed to Death: Rethinking Third World Development*. London: Green Print.

Turner, J. 1982. Issues in Self-Help and Self-Managed Housing. In: P. Ward (ed.), *Self-Help Housing: A Critique*, pp. 99–113. London: Alexandrine Press.

Turner, S. and Ingle, R. (eds) 1985. *New Developments in Nutrition Education*. Nutrition Education Series, vol. 11. Paris: UNESCO.

Turshen, M. 1989. *The Politics of Public Health*. New Brunswick, NJ: Rutgers University Press.

Twose, N. 1985. *Fighting the Famine*. London and San Francisco, Calif.: Pluto Press and Food First.

Tyler, C. 1990. Earthquake Zones – Sites of the World's Largest Cities. *Geographical Magazine* 62,3: 28–32.

Ul Haq, M. (ed.) 1991. *Human Development Report 1991*. New York: United Nations (UNDP).

Ullah, Md. S. 1988. Cyclonic-Surge Resistant Housing in Bangladesh: The Case of Urir Char. *Open House International* 13,2: 44–9.

UNDP (United Nations Development Programme) 1990a. *Human Development Report 1990*. New York: United Nations.

UNDP 1990b. *Disasters and Development – A Study of Institution Building*. Initial draft prepared by INTERTECT for UNDP April 1990.

UNDRO 1978. *Disaster Prevention and Mitigation: A Compendium of Current Knowledge*, vol. 2: *Hydrological Aspects*. New York: United Nations.

—— 1982a. *Disaster and the Disabled*. New York: United Nations.

—— 1982b. *Disaster Prevention and Mitigation: A Compendium of Current Knowledge*, vol. 8: *Sanitation Aspects*. New York: United Nations.

—— 1982c. *Shelter after Disaster – Guidelines for Assistance*. New York: United Nations.

—— 1984. *Disaster Prevention and Mitigation: A Compendium of Current Knowledge*, vol. 11: *Preparedness Aspects*. New York: United Nations.

—— 1985. The Day Before . . . . *UNDRO News* Nov.–Dec.: 4–6.

271

—— 1986. *Disaster Prevention and Mitigation: A Compendium of Current Knowledge*, vol. 12: *Social and Sociological Aspects*. New York: United Nations.

—— 1989. Editorial. *UNDRO News* May–June: 2.

—— 1991. *Mitigating Natural Disasters: Phenomena, Effects and Options – A Manual for Policy Makers and Planners*. Geneva: Office of the UN Disaster Relief Co-ordinator: 157.

UNICEF 1985. *Within Human Reach: A Future for Africa's Children*. New York: UNICEF.

—— 1988. *State of the World's Children 1988–89*. New York: Oxford University Press.

—— 1989. *Children on the Frontline: The Impact of Apartheid, Destabilization and Warfare on Children in Southern and South Africa*. New York: UNICEF.

—— 1992. *The State of the World's Children 1992*. New York: Oxford University Press.

Union News. 1988. Second Typhoon in Two Weeks Pounds Central Philippines. *Union News* 8, November: 3. Springfield, Mass.

United Nations 1980. Department of International, Economic and Social Affairs. *Demographic Yearbook*. New York: United Nations.

—— 1985. *Volcanic Emergency Management*. New York: Office of the United Nations Disaster Relief Co-ordinator.

—— 1986. *Urban and Rural Population Projections 1950– 2025: The 1984 Assessment*. New York: United Nations.

—— 1987. Logistics Key to Relief in Angola. *Africa Recovery* 4, December: 9–10.

United Nations Centre For Human Settlements (HABITAT) 1987. *Global Report on Human Settlements 1986*. Oxford: Oxford University Press.

United Nations Intergovernmental Panel on Climate Change 1990. *Study of Climate Change on Countries and Cities*. Report of Working Group 2. New York: United Nations.

van der Wusten, H. 1985. The Geography of Conflict since 1945. In: D. Pepper and A. Jenkins (eds), *The Geography of Peace and War*, pp. 13–28. Oxford: Basil Blackwell.

Vaughan, M. 1987. *The Story of an African Famine: Gender and Famine in Twentieth-Century Malawi*. Cambridge: Cambridge University Press.

Veltrop, J.A. 1990. Water, Dams and Civilisation. In: G. Le Moigne, S. Barghouti, and H. Plusquellec (eds), *Dam Safety and the Environment*, pp. 5–27. Washington, DC: World Bank.

Verhelst, T. 1990. *No Life Without Roots: Culture and Development*. London: Zed Press.

Verney, P. 1979. *The Earthquake Handbook*. New York and London: Paddington Press.

Vladut, T. 1990. Reservoir Induced Seismicity. In: G. Le Moigne, S. Barghouti, and H. Plusquellec (eds), *Dam Safety and the Environment*, pp. 113–26. Washington, DC: World Bank.

Walford, C. 1879. *Famines of the World: Past and Present*. New York: Burt Franklin.

Walgate, R. 1990. *Miracle or Menace? Biotechnology and the Third World*. Budapest: Panos Institute.

Walker, P. 1989. *Famine Early Warning Systems: Victims and Destitution*. London: Earthscan.

Ward, R. 1978. *Floods: A Geographical Perspective*. London: Macmillan.

Warmbrunn, W. 1972. *The Dutch Under German Occupation*. Stanford, Calif.: Stanford University Press.

Watts, M. 1983a. On the Poverty of Theory: Natural Hazards Research in Context. In: K. Hewitt (ed.), *Interpretations of Calamity*, pp. 231–62. Boston, Mass.: Allen & Unwin.

—— 1983b. *Silent Violence: Food, Famine and Peasantry in Northern Nigeria.* Berkeley, Calif.: University of California Press.

—— (ed.) 1986. *State, Oil and Agriculture in Nigeria.* Berkeley, Calif.: Institute of International Studies.

—— 1991. Entitlements or Empowerment? Famine and Starvation in Africa. *Review of African Political Economy* 51: 9–26.

WCED (World Commission on Environment and Development) 1987a. *Food 2000: Global Strategies for Sustainable Agriculture.* London: Zed Press.

WCED (World Commission on Environment and Development) (Brundtland Commission) 1987b. *Our Common Future.* Oxford: Oxford University Press.

Weir, D. 1987. *The Bhopal Syndrome.* London: Earthscan.

Wells, S. and Edwards, A. 1989. Gone with the Waves. *New Scientist* 11 November: 47–51.

Wescoat, J.L. 1991. The Flood Action Plan: A New Initiative Confronted by Basic Questions. *Natural Hazards Observer* 16,4: 1–2.

West, R. 1989. Richard West on Floods and Fear in Thailand. *The Independent Magazine* 28 October: 18.

Westergaard, K. 1983. *Pauperization and Rural Women in Bangladesh.* Dhaka: Bangladesh Academy for Rural Development.

Western, J.S. and Milne, G. 1979. Some Social Effects of a Natural Hazard: Darwin Residents and Cyclone Tracy. In: R.L. Heathcote and B.G. Thom (eds), *Natural Hazards in Australia*, pp. 488–502. Canberra: Australian Academy of Science.

Westing, A. (ed.) 1984a. *Environmental Warfare: A Technical, Legal, and Policy Appraisal.* London: Taylor & Francis.

—— (ed.) 1984b. *Herbicides in War: The Long-Term Ecological and Human Consequences.* London: Taylor & Francis.

—— (ed.) 1985. *Explosive Remnants of War: Mitigating the Environmental Effects.* London: Taylor & Francis.

Whitaker, J. 1988. *How Can Africa Survive?* New York: Harper & Row.

Whitcomb, G. 1990. Personal communications to Ian Davis from member of UNDRO disaster mitigation missions to Hong Kong.

White, A. 1981. *Community Participation in Water and Sanitation.* Technical Paper 17. Rijswijk, Netherlands: International Reference Centre for Community Water Supply and Sanitation.

White, A.U. 1974. Global Summary of Human Response to Natural Hazards: Tropical Cyclones. In: G.F. White (ed.), *Natural Hazards*, pp. 255–65. New York: Oxford University Press.

White, G. 1942. *Human Adjustment to Floods.* Research Paper 29. Chicago, Ill.: University of Chicago, Department of Geography.

—— 1973. Natural Hazard Research. In: J. Chorley (ed.), *Directions in Geography*, pp. 193–216. London: Methuen.

—— (ed.) 1974. *Natural Hazards.* New York: Oxford University Press.

White, S.C. 1988. In the Teeth of the Crocodile: Class and Gender in Rural Bangladesh. Ph.D. Dissertation, University of Bath.

Whittow, J. 1980. *Disaster: The Anatomy of Environmental Hazards.* Athens, Ga. and Harmondsworth: University of Georgia Press and Penguin.

WHO (World Health Organization) 1990. *Health For All When a Disaster Strikes*, vol. 2. Geneva: WHO/EPR.

Wijkman, A. and Timberlake, L. 1984. *Natural Disasters: Acts of God or Acts of Man?* London: Earthscan.

Wilches-Chaux, G. 1992a. The Global Vulnerability. In: Y. Aysan and I. Davis (eds), *Disasters and the Small Dwelling*, pp. 30–5. London: James & James Science Press.

— 1992b. Personal communication with Ian Davis regarding Nevado del Ruiz court case.

— 1993. Personal communication with Ian Davis regarding Galeras volcanic eruption of 15 January 1993.

Wilhite, D. and Easterling, W. (eds) 1987. *Planning for Drought: Toward a Reduction of Societal Vulnerability*. Boulder, Colo.: Westview.

Wilken, G. 1972. Microclimate Management by Traditional Farmers. *Geographical Review* 62: 544–60.

— 1977. Agroclimatic Hazards in Lesotho. Maseru, Lesotho and Boulder, Colo.: Ministry of Agriculture and Colorado State University, LASA mimeo.

— 1988. *Good Farmers*. Boulder, Colo.: Westview.

Wilkie, W.R. and Neal, A.B. 1979. Meteorological Features of Cyclone Tracy. In: R.L. Heathcote and B.G. Thom (eds), *Natural Hazards in Australia*, pp. 473–87. Canberra: Australian Academy of Science.

Williams, G. 1992. Personal communication to B. Wisner from Gene Williams, Graduate School of Education, University of Massachusetts, who had interviewed activists in Rio de Janeiro and São Paulo.

Wilmsen, E. 1989. *Land Filled with Flies: A Political Economy of the Kalahari*. Chicago, Ill.: University of Chicago Press.

Wilson, E.O. (ed.) 1988. *Biodiversity*. Washington, DC: National Academy Press.

— 1989. Threats to Biodiversity. *Scientific American* 261,3: 108–16.

Wilson, F. and Ramphele, M. 1989. *Uprooting Poverty: The South African Challenge*. New York: W.W. Norton.

Wilson, M. and Rachman, G. 1989. Are Hurricanes Growing in the Greenhouse? *Sunday Correspondent* (London) 24 September: 11.

Winchester, P. 1986. Cyclone Vulnerability and Housing Policy in the Krishna Delta, South India, 1977–83. Ph.D. Thesis, School of Development Studies, University of East Anglia.

— 1990. Economic Power and Response to Risk. In: J. Handmer and E. Penning-Rowsell (eds), *Hazards and the Communication of Risk*, pp. 95–110. Aldershot: Gower Publishing.

— 1992. *Power, Choice and Vulnerability: A Case Study in Disaster Mismanagement in South India, 1977–88*. London: James & James.

Wisner, B. 1975. An Example of Drought-Induced Settlement in Northern Kenya. In: I. Lewis (ed.), *Abaar: The Somali Drought*, pp. 24–5. London: International African Institute.

— 1976a. Health and the Geography of Wholeness. In: G. Knight and J. Newman (eds), *Contemporary Africa: Geography and Change*, pp. 81–100. Englewood Cliffs, NJ: Prentice-Hall.

— 1976b. Man-Made Famine in Eastern Kenya: The Interrelationship of Environment and Development. *Discussion Paper* 96, July. Brighton: Institute of Development Studies, University of Sussex.

— 1977. Constriction of a Livelihood System: The Peasants of Tharaka Division, Meru District, Kenya. *Economic Geography* 53,4: 353–7.

— 1978a. Letter to the Editor. *Disasters* 2,1: 80–2.

— 1978b. The Human Ecology of Drought in Eastern Kenya. Ph.D. Thesis, Graduate School of Geography, Clark University.

274

—— 1979. Flood Prevention and Mitigation in the People's Republic of Mozambique. *Disasters* 3,3: 293–306.

—— 1980. Nutritional Consequences of the Articulation of Capitalist and Non-Capitalist Modes of Production in Eastern Kenya. *Rural Africana* 8–9: 99–132.

—— 1982. Review of Dando's *The Geography of Famine. Progress in Human Geography* 6,2: 271–7.

—— 1984. Ecodevelopment and Ecofarming in Mozambique. In: B. Glaeser (ed.), *Ecodevelopment: Concepts, Projects, Strategies*, pp. 157–68. Oxford: Pergamon.

—— 1985. Natural Disasters. *Puerto Rico Libre* Autumn: 16–17.

—— 1987a. Doubts About Social Marketing. *Health Policy and Planning* 2,2: 178–9.

—— 1987b. Rural Energy and Poverty in Kenya and Lesotho: All Roads Lead to Ruin? *IDS Bulletin* 18,1: 23–9.

—— 1988a. GOBI vs. PHC: Some Dangers of Selective Primary Health Care. *Social Science and Medicine* 26,9: 963–9.

—— 1988b. *Power and Need in Africa: Basic Human Needs and Development Policies*. London and Trenton, NJ: Earthscan and Africa World Press.

—— 1990. Harvest of Sustainability: Recent Books on Environmental Management. *Journal of Development Studies* 26,2: 335–41.

—— 1992a. Health of the Future/The Future of Health. In: A. Seidman and F. Anang (eds), *21st Century Africa: Towards a New Vision of Self-Sustainable Development*, pp. 149–81. Trenton, NJ and Atlanta, Ga.: Africa World Press and African Studies Association.

—— 1992b. Too Little to Live On, Too Much to Die From: Lesotho's Agrarian Options in the Year 2000. In: A. Seidman, K. Mwanza, N. Simelane, and D. Weiner (eds), *Transforming Southern African Agriculture*, pp. 87–104. Trenton, NJ: Africa World Press.

Wisner, B., Gilgen, H., Antille, N., Sulzer, P., and Steiner, D. 1987. A Matrix-Flow Approach to Rural Domestic Energy: A Kenyan Case Study. In: C. Cocklin, B. Smit, and T. Johnston (eds), *Demands on Rural Lands: Planning for Resource Use*, pp. 211–39. Boulder, Colo.: Westview.

Wisner, B. and Mbithi, P. 1974. Drought in Eastern Kenya: Nutritional Status and Farmer Activity. In: G. White (ed.), *Natural Hazards*, pp. 87–97. New York: Oxford University Press.

Wisner, B., O'Keefe, P., and Westgate, K. 1977. Global Systems and Local Disasters: The Untapped Potential of People's Science. *Disasters* 1,1: 47–57.

Wisner, B., Stea, D., and Kruks, S. 1991. Participatory and Action Research Methods. In: E. Zube and G. Moore (eds), *Advances in Environment, Behavior, and Design*, vol. 3, pp. 271–95. New York: Plenum.

Wisner, B., Westgate, K., and O'Keefe, P. 1976. Poverty and Disaster. *New Society* 9 September: 547–8.

Woldermariam, M. 1984. *Rural Vulnerability to Famine in Ethiopia: 1958–1977*. Addis Ababa: Vikas.

Wood, R.M. 1986. *Earthquakes and Volcanoes*. London: Mitchell Beazley.

Woodham-Smith, C. 1962 *The Great Hunger: Ireland 1845–9*. London: Hamish Hamilton.

Woodruff, B.A., Toole, M.J., Rodrigue, C., Brink, E.W., El Sadig Mahgoub, Magda Mohamed Ahmed, and Babikar, A. 1990. Disease Surveillance and Control After a Flood: Khartoum, Sudan 1988. *Disasters* 14,2: 151–63.

World Bank 1989. *Sudan Forestry Sector Review*. Washington, DC: World Bank.

—— 1990. Flood Control in Bangladesh: A Plan for Action. Asia Region Technical

Department. World Bank Technical Paper No. 119. Washington, DC. World Bank.

World Resources Institute 1986. *World Resources 1986*. New York: Basic Books.

—— 1988. *World Resources 1988-89*. New York: Basic Books.

York, S. 1985. Report on a Pilot Project to Set Up a Drought Information Network in Conjunction with the Red Crescent Society in Darfur. *Disasters* 9,3: 173–9.

Young, K. (ed.) 1988. *Women and Economic Development*. Oxford and Paris: Berg and UNESCO.

Zaman, M.Q. 1988. The Socioeconomic and Political Dynamics of Adjustment to Riverbank Erosion Hazard and Population Resettlement in the Brahmaputra–Jamuna Floodplain. D.Phil. Dissertation, University of Manitoba.

—— 1991. Social Structure and Process in Char Land Settlement in the Brahmaputra–Jamuna Floodplain. *Man* 26,4: 549–66.

Zarco, R.M. 1985. *Anticipation, Reaction and Consequences: A Case Study of the Mayon Volcano Eruption*. Quezon City: Philippine Institute of Volcanology and Seismology Research Staff.

Zeigler, D.J., Johnson, J.H., and Brunn, S.D. 1983. *Technological Hazards*. Resource Publications in Geography. Washington, DC: Association of American Geographers.

Zhao, S. 1986. *Physical Geography of China*. Chichester: John Wiley.

Ziegler, P. 1970. *The Black Death*. London: Pelican.

INDEX

Page numbers in **bold** type refer to figures and tables.